AA003392

American Institute of Chemical Engineers

# Conference of the Nuclear Engineering Division 2007

Held at the 2007 AIChE Spring National Meeting

April 22-27, 2007
Houston, Texas, USA

**Printed from e-media with permission by:**

Curran Associates, Inc.
57 Morehouse Lane
Red Hook, NY 12571
www.proceedings.com

**ISBN: 978-1-60423-365-0**

**Some format issues inherent in the e-media version may also appear in this print version.**

American Institute of Chemical Engineers

# Conference of the Nuclear Engineering Division 2007

Held at the 2007 AIChE Spring National Meeting

AIChE

April 22-27, 2007

Houston, Texas, USA

ISBN: 978-1-60423-365-0



American Institute of Chemical Engineers
Proceedings
Three Park Avenue
New York, NY 10016-5991
Phone: 212-591-8100

www.aiche.org

American Institute of Chemical Engineers

Conference of the Nuclear Engineering Division
2007

# TABLE OF CONTENTS

## Session 2 – Advanced Oxidation and Separation Processes for Environmental & Energy Industries



## Session 33 – Hydrogen Production from Nuclear Energy

# Purification of Simulated Neptunium Filtrate Solution by Anion Exchange

**Mark L. Crowder** and Edward A. Kyser, III. Savannah River National Laboratory, Washington Savannah River Company, Aiken, SC 29808

Anion exchange was investigated as a means to recover and purify neptunium (Np) from solutions containing significant plutonium (Pu), sodium (Na), and manganese (Mn) impurities and a high nitrate concentration. The solution being evaluated resulted from lean filtrate and precipitator cleanout solutions which were generated during Np processing at the Savannah River Site. Prior to anion exchange, valence adjustment of the Np using ferrous sulfamate (i.e., Fe2(SO3NH2)2 or FS) was evaluated. For testing, a simulated filtrate solution was made with ~1.5 g Np/L and slightly elevated levels of impurities. Two anion exchange column runs were performed. In both runs, Np was loaded onto Reillex HPQ resin in high nitric acid (HNO3) solution, typical for Np processing. The resin wash step differed in the two runs, but the elution step was the same. In the first column run, 12 column bed volumes (BV) of reductive or partition wash with 6.4 M nitric acid / 0.05 M FS were used. In the second run, six BV of 5 M HNO3 / 0.05 M FS were used as the reductive wash. Reported results will include Pu rejection, Np losses and decontamination factors for Na and Mn.

## Cesium Removal Performance of Resorcinol-Formaldehyde Resin

**Charles A. Nash**[1], Sebastian E. Aleman[2], Cheryl E. Duffey[1], and Murray R. Thorson[3]. (1) Chemical Science & Technology, Savannah River National Laboratory, Savannah River Site, Building 773-42A, room 182, Aiken, SC 29808, (2) Engineering, Modeling and Simulation, Savannah River National Laboratory, Savannah River Site, Building 773-42A, room 182, Aiken, SC 29808, (3) Hanford Waste Treatment Plant, Richland, WA 99352

The Department of Energy's Hanford River Protection Project is an effort to design and construct a Waste Treatment Plant (WTP) for processing and stabilization of nuclear waste from past operations at Hanford, WA. As part of that effort the Savannah River National Laboratory (SRNL) has been contracted to test technology for unit operations including the use of ion exchange to remove cesium ion from highly alkaline aqueous feed.

Resorcinol-formaldehyde (RF) cationic exchange resin is highly selective for cesium under alkaline conditions. The work presented here examined the isotherm, capacity, kinetics, and column performance of a spherical form of RF. The spherical or bead form of the resin has been found to be hydraulically better than the traditional ground gel form of RF in past work, and the different method of manufacture necessitated the study of the resulting different chemical performance.

The resin isotherm was found to depend primarily on sodium, potassium, and free hydroxide levels along with temperature. In addition the resin was found to have a high content (6 +/- 0.8 meq/gram hydrogen form) of non-selective cationic sites along with roughly 0.8 meq/gram of sites that select cesium under WTP feed compositions. Standard batch contact methods were used to obtain data for isotherm modeling. Loaded resin samples were also analyzed to determine capacity and also affinities for other elements in the feeds. Chromium, iron, lead, and cadmium were found to be adsorbed by the resin while many other metals in the feed are not selected.

Kinetics testing was performed to assist in modeling resin performance under dynamic conditions. Kinetics data are included in the modeling of column performance. Small columns were run to breakthrough to provide data for benchmarking.

The presentation will include comparisons of isotherm models with batch data. SRNL uses a combination of Pitzer solution chemical activity modeling and an algebraic model to provide a cesium isotherm.

This work shows that RF resin is a viable choice for the cesium removal process at WTP. The extensive chemical characterizations of the resin provide a good basis for predictions of process performance under a wide range of Hanford feed compositions.

# Appropriateness of mechanistic and non-mechanistic models for the application of ultrafiltration to mixed waste

By

Henry Foust, Ph.D.
Malay Ghosehajra, Ph.D., P.E.

### *Abstract*

This study asks two questions: (1) How appropriate is the use of a basic filtration equation to the application of ultrafiltration of mixed waste, and (2) How appropriate are non-parametric models for permeate rates (volumes)? To answer these questions, mechanistic and non-mechanistic approaches are developed for permeate rates and volumes associated with an ultrafiltration/mixed waste system in diafiltration mode. The mechanistic approach is based on a filtration equation which states that t/V vs. V is a linear relationship. The coefficients associated with this linear regression are composed of physical/chemical parameters of the system and based the mass balance equation associated with the membrane and associated developing cake layer. For several sets of data, a high correlation is shown that supports the assertion that t/V vs. V is a linear relationship. It is also shown that non-mechanistic approaches, i.e., the use of regression models to are not appropriate. One models considered is Q(p) = a*ln(Cb)+b. Regression models are inappropriate because the scale-up from a bench scale (pilot scale) study to full-scale for permeate rates (volumes) is not simply the ratio of the two membrane surface areas.

### *Introduction*

It is often the case without a full understanding of some phenomena that scientist resort to regression models to express relationships between states of the system of concern. An example of this approach is the application of ultrafiltration to mixed waste. Much is not understood in this system; one example is that the states that influence permeate rates are not well understood. Often, a bench (pilot) scale study is conducted to generate data of $C_b$ (time) vs. permeate rates that can be used for a full-scale process. From this data, a regression model is developed. It has been this authors experience that one model is

(Equation 1.)

$$Q_p = a\ln(C_b) + b$$

Where {a, b} are regression parameters, $C_b$ is the bulk phase concentration, and $Q_p$ is the permeate rate. A caution associated with any regression model is that it is only as good as the data it is generated from and should not be used for prediction out of sample.

This caution is important, because it has been observed that regression models derived from bench (pilot) scale results are often scaled to predict the performance of a full-scale process, such as ultrafiltration. The scaling is simply a ratio of the two membrane surface areas for the bench (pilot) scale and full-scale apparatus. One example of this is that an ultrafiltration process (UFP) is being designed and built at the Hanford Department of Energy facility. Many bench scale experiments have been conducted to support the design of the UFP and regression models have been utilized to model dewatering experiments. These models have also been extrapolated to the full-scale UFP to determine production levels, but is this method based on sound theory?

This study will ask three questions: (1) how appropriate is the use of a basic filtration equation to the application of ultrafiltration of mixed waste? (2) Why are coefficients associated with the mechanistic model time invariant when the physical parameters within this coefficient are time varying? (3) How appropriate are non-parametric models for permeate rates (volumes)?

### *Background*

Using Darcy's law (Geankoplis, 2003), the following relationship can be established

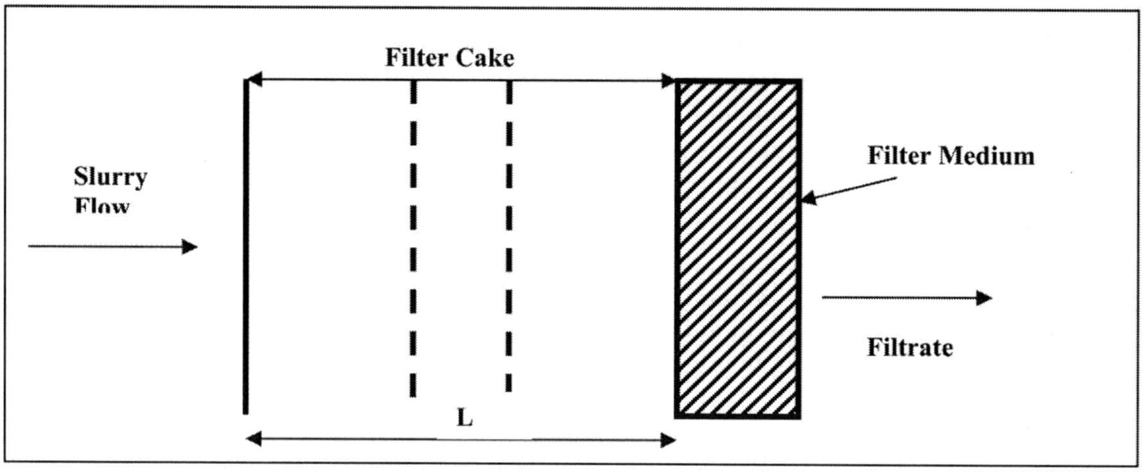


**Figure 1 – Developing Cake Layer**

**Figure 1** is a section through a filter cake and filter medium at a definite time 't' from the start of the flow of filtrate. This flow can be described by an equation similar to Poiseuille's law as follows

(Equation 2.)

$$\frac{\Delta P}{L} = \frac{32\mu v}{gD^2}$$

For laminar flow in a packed bed of particles, the Carman-Kozeny relation, which is similar to **Equation 2** has been shown to apply to filtration as follows

(Equation 3.)

$$\frac{\Delta p_c}{L} = \frac{K_1 \mu v (1-\varepsilon)^2 S_0^2}{\varepsilon^2}$$

The linear velocity is based on the empty cross-sectional area and is given by

(Equation 4.)

$$v = \frac{\frac{dV}{dt}}{A}$$

The cake thickness, L, may be related to the volume of filtrate, V, by a material balance. If 'C$_s$" is mass of filtrate, a material balance gives

(Equation 5.)

$$LA(1-\varepsilon)p_p = C_s(V + \varepsilon LA))$$

Where, $p_p$ is the density of solid particles in the cake.

Neglecting the final term in **Equation 5** and substituting **Equation 4** into **Equation 3**, we obtain

(Equation 6.)

$$\frac{dV}{Adt} = \frac{\Delta p_c}{\frac{K_1(1-\varepsilon)S_0^2}{p_p\varepsilon^3}\frac{\mu C_s V}{A}} = \frac{\Delta p_c}{\alpha\frac{\mu C_s V}{A}}$$

Where Specific Cake Resistance is denoted by $\alpha = \frac{K_1(1-\varepsilon)S_0^2}{p_p\varepsilon^3}$

For the filter medium resistance, it can be written, by analogy with **Equation 6**

(Equation 7.)

$$\frac{dV}{Adt} = \frac{\Delta p_f}{\mu R_m}$$

Where $R_m$ = Resistance of filter medium to filtrate flow

Since the resistance of the cake and the filter medium are in series, **Equations 4** and **7** can be combined and becomes

(Equation 8.)

$$\frac{dV}{Adt} = \frac{\Delta p}{\mu\left(\frac{\alpha C_s V}{A} + R_m\right)}$$

Where $\Delta p = \Delta p_c + \Delta p_f$

Often a filtration occurs under constant pressure condition, therefore, **Equation 8** can be inverted and rearranged to give

(Equation 9.)

$$\frac{dt}{dV} = \frac{\mu \alpha C_s}{A^2 \Delta p} V + \frac{\mu}{A \Delta p} Rm$$

For constant pressure, constant α, and incompressible cake, *V* and *t* are the only variables in **Equation 9**. Integrating to obtain the time of filtration in 't' sec we obtain

(Equation 10.)

$$t = aV^2 + bV$$

Where

$$a = \frac{\mu \alpha Cs}{2A^2 \Delta p}, b = \frac{\mu R_m}{A \Delta p}$$

Where, $\mu$ = viscosity of filtrate

$$\alpha = \frac{K_1(1-\varepsilon)S_0^2}{p_p \varepsilon^3}$$, known as Specific Cake Resistance

$C_s$ = Mass of filtrate
$A$ = Filter Area
$\Delta p$ = Pressure Drop
$R_m$ = Resistance of filter medium to filtrate flow
$\varepsilon$ = Void fraction or porosity of cake
$p_p$ = density of solid particles in Cake
$S_o$ = Specific Surface Area of Particle per volume of solid particle

At this point in the research, it was asked how appropriate is **Equation 2** when it is known that $C_b$, α and Δp change with time? To answer this question, empirical data was utilized to model t/V vs. V. If a linear fit is appropriate, then there is evidence that **Equation 10** is appropriate.

Given in **Figure 2** is the fit of t/V vs. V for a set of experimental results from Duignan et al. (2003). As can be seen from the figure, there is a high correlation between t/V and V, which supports the above assertion. Other data sets will be explored in the next section.

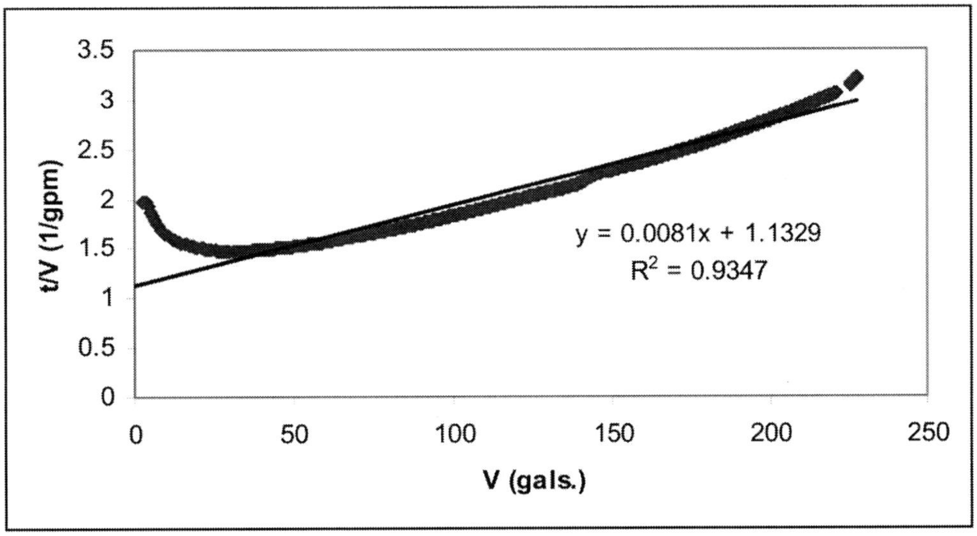


**Figure 2 – V vs. t/V for Duignan**

Another approach to modeling this data is simple to use a regression model. Two regression models are often utilized to predict the behavior of permeate rates. These models are

(Equation 11.)

$$Q_p = Q_0 e^{-kt}$$

*or*

$$Q_p = a\ln(C_b) + b$$

As can be seen in **Figure 3**, both a mechanistic model and non-mechanistic model work well for the given empirical data. The problem lies when we attempt to scale the results of a model that is based on a non-parametric method to another scale.

### *Model Validation*

To validate the appropriateness of treating t/V vs. V as a linear relationship, several data sets associated with the UFP at Hanford have been modeled. Many data sets do exist, but the study conducted by Duignan et al. (2003) is most representative of a full-scale UFP in terms of mechanical scale issues. This effort is summarized in **Table 1**. A high correlation is shown for all data sets and this supports the assertion that coefficients {a, b} are constant for most of the dewatering phase.

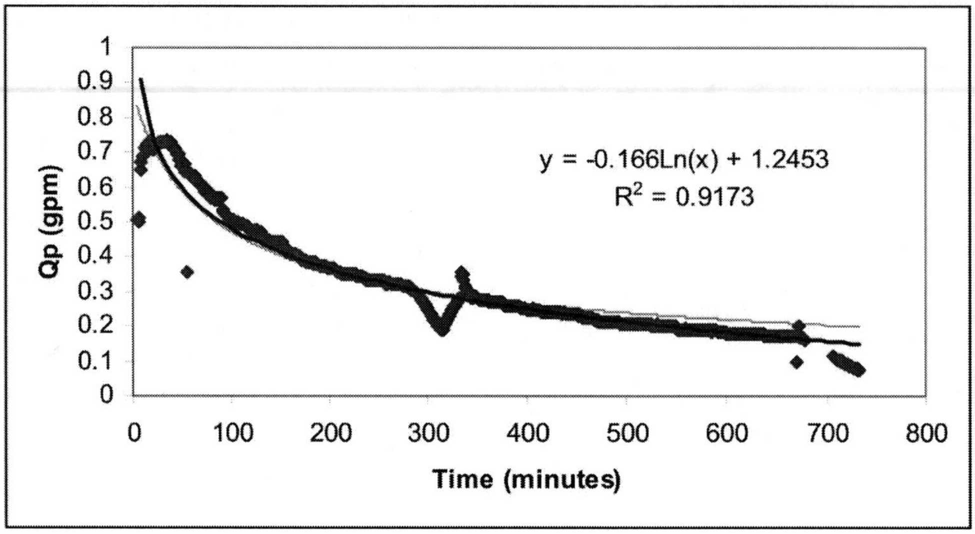


**Figure 3 – t vs. Qp**

| Data Set | a | b | $R^2$ |
|---|---|---|---|
| 3A | .0264 | 1.7741 | .9906 |
| 3B | .0081 | 1.1329 | .9347 |
| 3C | .0546 | 1.7878 | .9719 |
| 4A | .0192 | 4.8083 | .9381 |

**Table 1 – Data Sets**

### *Non-Mechanistic Models*

In this section, it will be demonstrated that these models are not appropriate when utilized to scale results from a bench (pilot) scale study to a full-scale study. **Table 2** provides the values for the mechanistic model at the scale seen in **Figure 3** and **Figure 4** represents the scaling of the mechanistic model and non-mechanistic model to a larger scale scaled by the factor (200/6.7), which is roughly 30. As can be seen in **Figure 4** (where the mechanistic model remains above the x-axis), there is a great deal of discrepancy between the results from both models and the non-mechanistic approach appears to predict negative permeate rates; this is not physically possible.

| Parameters for A = 6.7 ft^2 | Parameters for A = 200 ft^2 |
|---|---|
| a = .0081 | a = 10E-6 |
| b = 1.1329 | b = .04 |

**Table 2 – Scale Issues**

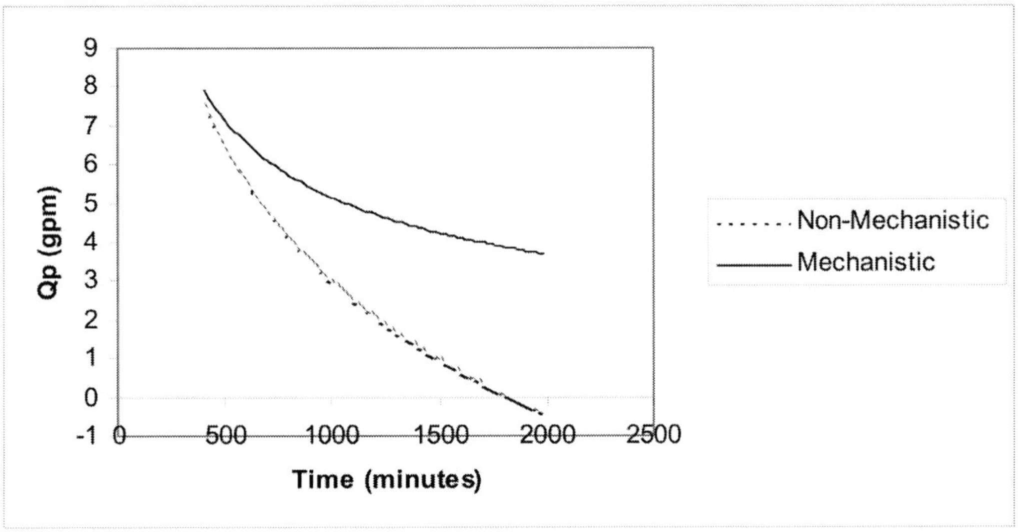


**Figure 4 – T vs. Qp**

## *Discussion*

It has been shown that a linear relationship exists between t/V vs. V for radioactive waste slurry treated by ultrafiltration. **Table 2** has shown for several data sets a high correlation ($R^2$ >.90) exists. This model is more appropriate than non-mechanistic (regression) models, because it accounts for physical scale effects.

The linear model for t/V vs. V includes the parameters {a, b} which are based on several physical/chemical parameters that change with the dewatering processes that occur and it is asked how can the parameters {a, b} remain fixed when they are composed of parameters that are time varying? No apparent explanation was found.

Data from Duignan et al. (2003) was utilized, because this data set appears to best deal with issues of scale. In Duignan et. al (2003), seven membrane tubes are bundled together to give more representative fluid dynamics; the membrane tube length is closer to the membrane tube length of the full-scale model; and the results were from dewatering experiments over the complete range of $C_b$.

Regression models derived from bench-scale results fit quite well to the data and are useful to predict Qp or Vp when utilized in sample and at the same scale. But are not appropriate out of sample or a different scale, i.e., a different membrane surface area.

## *Conclusions*

Regression models to predict t vs. Qp or t vs. Vp where t is time and Qp and Vp are permeate rates or volumes are only appropriate when used within sample at the same scale as the utilized bench-scale results. When regression models of the form

(Equation 12.)

$$Q_p = Q_0 e^{-kt}$$

*or*

$$Q_p = a\ln(C_b) + b$$

Are developed for bench scale results and an attempt is made to scale these results to a full-scale process, then results may be seen as in **Figure 4** and are surely not representative of the process. Another approach is to utilize

(Equation 13.)

$$\frac{t}{V} = aV + b$$

Which physically accounts for changes in membrane surface area and other scaling of other parameters.

It is also noted that the results of this paper support (Foust, 2006) an upcoming paper that utilizes the physically derived model for permeate rates and a mass balance equation for $C_b$ to determine dewatering times as a function of V, A, and C* where

(Equation 14.)

$$C^* = \left(\frac{C_f}{C_0} - 1\right)$$

### *References*

Duignan, Mark (2003). "Final Report: Pilot-Scale Cross-flow ultrafiltration test using a Hanford Site Tank 241-AN-102 Waste Simulant (U)." Westinghouse Savannah River Company report, SRT-RPP-00087, Rev. 0.

Foust, Henry (2006). "Minimizing membrane surface area for a given diafiltration problem." American Society of Civil Engineers, *Journal of Environmental Engineering* (submitted).

Geankoplis, C.J. (2003) "Transport processes and separation process principles." Prentice-Hall, Upper Saddle River, NJ.

## Yield Stress Reduction of DWPF Melter Feed Slurries

*M. E. Stone*
*M. E. Smith*
*Savannah River National Laboratory, Aiken, SC 29808*

The Defense Waste Processing Facility (DWPF) at the Savannah River Site vitrifies High Level Waste for repository internment. The process consists of three major steps: waste pretreatment, vitrification, and canister decontamination/sealing. The HLW consists of insoluble metal hydroxides (primarily iron, aluminum, magnesium, manganese, and uranium) and soluble sodium salts (carbonate, hydroxide, nitrite, nitrate, sulfate). The pretreatment process acidifies the sludge with nitric and formic acids, adds the glass formers as glass frit, then concentrates the resulting slurry to approximately 50 weight percent (wt%) total solids. This slurry is fed to the joule-heated melter where the remaining water is evaporated followed by calcination of the solids and conversion to glass.

The Savannah River National Laboratory (SRNL) is currently assisting DWPF efforts to increase throughput of the melter. As part of this effort, SRNL has investigated methods to increase the solids content of the melter feed to reduce the heat load required to complete the evaporation of water and allow more of the energy available to calcine and vitrify the waste. The process equipment in the facility is fixed and cannot process materials with high yield stresses, therefore increasing the solids content will require that the yield stress of the melter feed slurries be reduced.

Changing the glass former added during pretreatment from an irregularly shaped glass frit to nearly spherical beads was evaluated. The evaluation required a systems approach which included evaluations of the effectiveness of beads in reducing the melter feed yield stress as well as evaluations of the processing impacts of changing the frit morphology. Processing impacts of beads include changing the settling rate of the glass former (which effects mixing and sampling of the melter feed slurry and the frit addition equipment) as well as impacts on the melt behavior due to decreased surface area of the beads versus frit.

Beads were produced from the DWPF process frit by fire polishing. The frit was allowed to free fall through a flame, then quenched with a water spray. Approximately 90% of the frit was converted to beads by this process, as shown in Figure 1. Borosilicate beads of various diameters were also procured for initial testing.

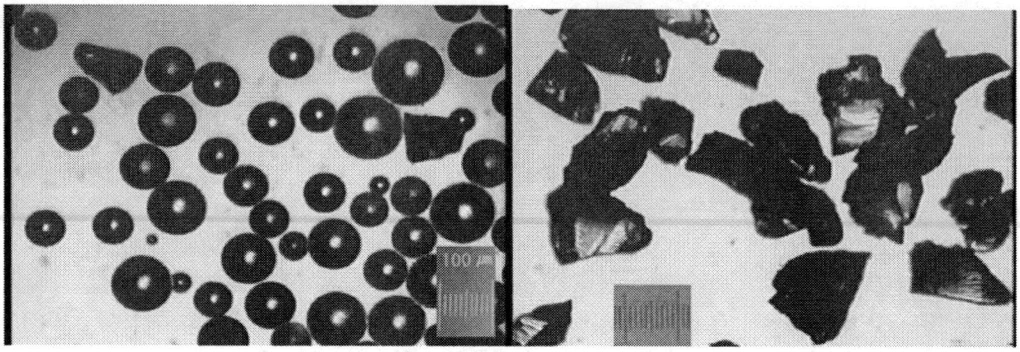

**Figure 1. Beads and Unprocessed Frit 418**

Yield stress reduction was measured by preparing melter feed slurries (using non-radioactive HLW simulants) that contain beads and comparing the yield stress with melter feed containing frit. The yield stress was measured with a Haake RS-150 or Haake RS-600 rheometer. The Bingham Plastic model was used to regress the data and determine the yield stress. Results of preliminary tests are shown in Table 1, with the beads resulting in a decrease of 20 to 40%[1].

**Table 1. Preliminary Results: Beads versus Frit**

| SB2 | Yield Stress | Consistency | SB2/3 | Yield Stress | Consistency |
|---|---|---|---|---|---|
| | 50 - 200 Shear Rate Up Curve Only | | | 50 - 200 Shear Rate Up Curve Only | |
| Beads | Pa | cP | Beads | Pa | cP |
| RHEO-226-A | 4.1 | 14.6 | RHEO-229-A | 2.8 | 16.5 |
| RHEO-226-B | 4.4 | 15.3 | RHEO-229-B | 3.0 | 17.3 |
| Average | 4.2 | 15.0 | Average | 2.9 | 16.9 |
| Frit | Pa | cP | Frit | Pa | cP |
| RHEO-227-A | 5.7 | 18.1 | RHEO-228-A | 4.5 | 20.5 |
| RHEO-227-B | 5.0 | 13.9 | RHEO-228-B | 5.5 | 21.6 |
| Average | 5.4 | 16.0 | Average | 5.0 | 21.0 |
| % Difference | 21.4 | 6.1 | % Difference | 41.5 | 19.7 |

A second set of tests was performed with beads of various diameters to determine if a decrease in diameter affected the results. Smaller particle size was shown to increase yield stress when frit is utilized. The settling rate of the beads was required to match the settling rate of the frit, therefore a decrease in particle size was anticipated. As shown in Table 2, decreasing the particle size did not reduce the effectiveness of switching from frit to beads[2].

**Table 2. Impact of Frit 320 and Various Sized Beads on Rheology of SB3 SME Product**

| Frit/Bead (Sample ID) | Yield Stress | Consistency (Pa) | * % Reduction in Yield Stress /Consistency |
|---|---|---|---|
| Frit 320 | 3.26 | 20.53 | N.A. |
| -140 Bead | 2.64 | 16.05 | 23 / 28 |
| -100+140 Bead | 2.51 | 15.27 | 30 / 34 |
| -70+100 Bead | 2.61 | 12.88 | 25 / 59 |

* Relative to Frit 320 SME product

Settling tests were conducted in water, xanthan gum solutions, and in non-radioactive simulants of the HLW. The tests used time-lapse videography as well as solids sampling to evaluate the settling characteristics of beads compared to frit of the same particle size. The test with HLW simulant was performed by pouring a well-mixed 45% slurry into a 100 ml cylinder and sampling the each cylinder at the 50 ml volume for total solids as a function of time. As shown in Table 3, beads settled slightly faster than frit at a given particle size.

**Table 3. Settling Tests with SB3 Melter Feed**

| Settling Time (Hours) | Wt% Solids of SB3/Frit 320 (-100+140 mesh) | Wt% Solids of SB3/Beads (-100+140 mesh) |
|---|---|---|
| 2 | 46.52 | 47.17 |
| 4 | 45.79 | 48.10 |
| 6 | 45.67 | 47.36 |
| 24 | 45.95 | 49.26 |

Settling rate for larger frit particles (-80/+100 mesh) was much higher than the settling rate for the -100/+140 beads, so the particle size range for beads was reduced from the -80/+200 specified for frit to -100/+200 for the beads.

A preliminary melt rate evaluation was performed using a dry-fed Melt Rate Furnace (MRF) developed by SRNL. The MRF is a scoping tool designed for quick screening of melt rate. The testing consisted of preparing sufficient melter feed simulant using beads and frit to produce 500 grams of glass. This material is dried and placed into a 1200 ml stainless steel beaker. The beaker is heated from the bottom with the sides insulated for a specified time, then the amount of glass in the beaker is measured. The measured melt rates (0.45 and 0.47 in/hr) were nearly identical, but differences in the appearance of the fired beakers were noted as shown in Figure 2.

**Figure 2. Top Views of the SB3/Bead 320 and SB3/Frit 320 MRF Beakers and the Cross-Sections of the Beakers**

The differences in the appearances of the beakers in MRF indicate that further evaluation is necessary. Tests are planned in the Slurry-fed Melt Rate Furnace that more closely replicates the conditions in the DWPF melter.

Preliminary evaluation of the impact of beading the frit on the frit addition system were completed by conducting flow loop testing. A recirculation loop made of clear ¾" PVC piping was built with a total length of about 30 feet. The test rig had a plastic 10 gallon feed tank and a Jabsco pump. A 2" PVC bypass return loop to the feed tank was installed immediately after the pump with a ball valve to control flow in the main recirculation loop. A flow transmitter followed by both a pressure gauge and a digital pressure gauge were installed on the outlet side of the pump. A schematic of the bead pump test stand and a picture of the test stand (without the bypass line) are shown in Figures 3 and 4 respectively. Six gallon solutions of water and either -100+140 mesh beads or Frit 320 were tested. In the first test, 14 kg of water was first added to the feed tank, and then 14.0 kg of beads (to make a 50 weight percent solids solution) was slowly added with the agitator running. A minimum agitator speed was then determined. After this, the pump was started with bypass valve closed for maximum recirculation flow. Pump power, flow rate, outlet pressure, and observations of the flow in the horizontal upper section of the loop were noted. The recirculation flow was then gradually reduced and the above items recorded until settling was noted in the recirculation line. After this, the test was repeated by adding 3.11 kg of beads to make a 55 weight percent mixture. Finally, a 60 weight percent mixture was tested by adding another 3.89 kg of beads to the feed tank. This same process was planned for Frit 320, but only the initial 50 weight percent mixture could be tested before the pump impeller failed due to the erosive Frit 320 particles.

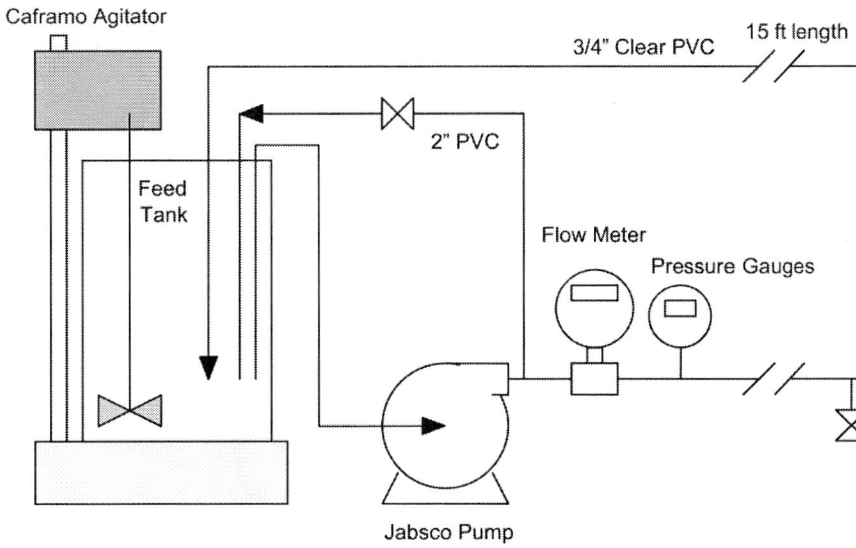


**Figure 3. Schematic of Bead Pump Test Stand**

**Figure 4. Bead Pump Test Stand**

Overall, the data shows that the line pressure increased as the solids were increased for the same flow rate. In addition, the line pressure was higher for Frit 320 than the beads at the same solids level and flow. With the observations, a determination of minimum velocity to prevent settling could be done, but a graph of the line pressures (see Figure 5) versus velocity (not flow) for the various tests was deemed to more objective. The graph shows that the inflection point in pressure drop is about the same for the beads and Frit 320 (about 2.5 ft/sec). This indicates that the bead (-100+140 mesh) slurry would not require higher flows rates than frit slurry at DWPF during transfers. Another key finding was that the pump impeller was not significantly damaged by the bead slurry, while the Frit 320 slurry rapidly destroyed the impeller. Evidence of this was first observed when black particles were seen in the Frit 320 slurry being recirculated and then confirmed by a post-test inspection of the impeller. Finally, the pumping of bead slurry could be recovered even if flow is stopped. The Frit 320 slurry could not be restarted after stopping flow due to the nature of the frit to pack tightly when settled.

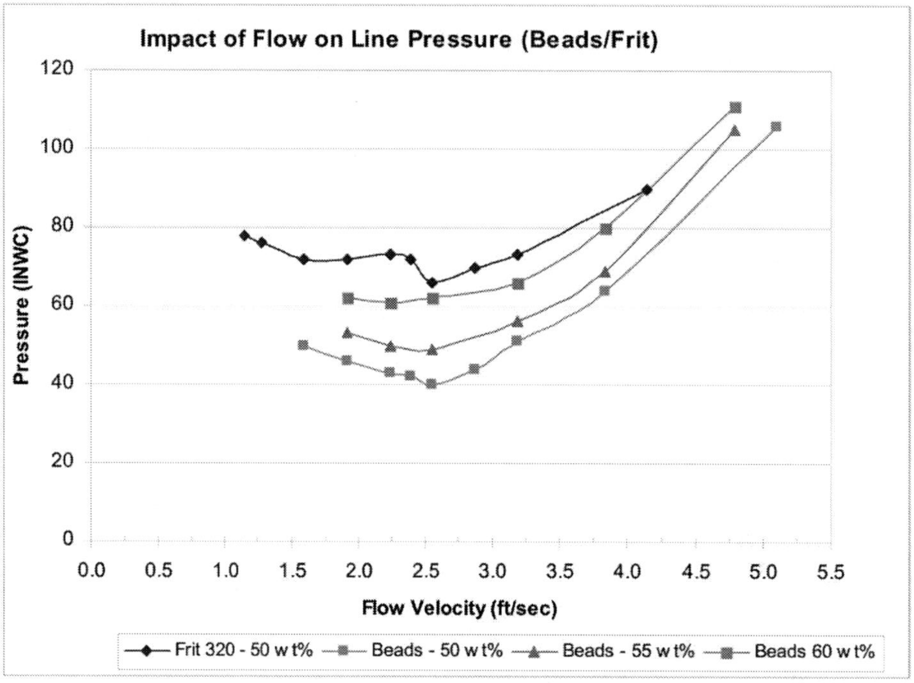


**Figure 5. Line Pressure versus Velocity**

Beads were shown to represent a significant process improvement versus frit for the DWPF process in lowering yield stress of the melter feed. Lower erosion of process equipment is another expected benefit.

Additional experimental work planned in the following areas:

- SME product homogeneity/sampling issues
- Melt rate during slurry-fed tests
- Impact of beads on frit addition equipment
    - Individual component evaluation
- Testing with various ratios of frit to beads

References

[1] Stone, M. E., Schumacher, R. F., 2004,"Preliminary Evaluation of Glass Former Morphology Impacts on Rheological Properties of Simulated DWPF Melter Feed Slurries", WSRC-TR-2004-00337, Washington Savannah River Company, Aiken, SC.

[2] Smith, M.E., Stone, M. E. Stone, D. H. Miller, 2005,"Impact of Spherical Frit Beads on Simulated DWPF Fluids" WSRC-TR-2005-00148, Washington Savannah River Company, Aiken, SC.

# COMBINING ADVANCED OXIDATION PROCESSES: ASSESSMENT OF PROCESS ADDIVITIVITY, SYNERGISM, AND ANTAGONISM

***ROBERT W. PETERS*, Ph.D., P.E.**
*Professor of Environmental Engineering*
*Department of Civil, Construction, and Environmental Engineering*
*University of Alabama at Birmingham*
*1075 13th Street South*
*Birmingham, Alabama*

***M.P. SHARMA, Ph.D.***
*Professor of Chemical Engineering*
*Department of Chemical and Petroleum Engineering*
*University of Wyoming*
*Laramie, Wyoming*

***YUSUF GBADEBO ADEWUYI, Ph.D.***
*Professor of Chemical Engineering*
*Department of Mechanical and Chemical Engineering*
*North Carolina Agricultural and Technical State University*
*1601 East Market Street, 326 McNair Hall*
Greensboro, North Carolina

## Abstract

This paper addresses the process interactions from combining integrated processes (such as advanced oxidation processes (AOPs), biological operations, air stripping, etc.). AOPs considered include: Fenton's reagent, ultraviolet light, titanium dioxide, ozone ($O_3$), hydrogen peroxide ($H_2O_2$), sonication/acoustic cavitation, among others. A critical review of the technical literature has been performed, and the data has been analyzed in terms of the processes being additive, synergistic, or antagonistic. Predictions based on the individual unit operations are made and compared against the behavior of the combined unit operations. The data reported in this paper focus primarily on treatment of petroleum hydrocarbons and chlorinated solvents.

# Introduction

### *Description of Advanced Oxidation Processes*

Acoustic cavitation (commonly termed sonication) involves the application of sound waves being transmitted through a liquid as a wave of alternating cavitation cycles. Compression cycles exert a positive pressure on the liquid, pushing molecules together, while expansion cycles exert a negative pressure, pulling molecules away from each other. The chemical effects of sonication are a result of acoustic cavitation. During rarefaction, molecules are torn apart forming tiny microbubbles that grow to a critical size during the alternating cavitation cycles, and then implode releasing a large amount of energy. Temperatures on the order of 5,000°K and pressures up to 500 to 1,000 atmospheres have been observed in microbubble implosions, while the bulk solution stays near ambient. The collapsing bubble interface results in the formation of hydroxyl and hydrogen radicals. These radicals destroy chlorinated organic and petroleum hydrocarbon compounds very effectively.

The performance can be further enhanced through the addition of advanced oxidants (e.g., ozone ($O_3$), hydrogen peroxide ($H_2O_2$), etc.), and incorporation of recent advancements in the acoustic cavitation field [1]. The destruction of organic pollutants can occur via several mechanisms. The organic pollutant inside the cavity and in the interfacial region can undergo pyrolysis reactions (or combustion reactions if oxygen is present) during the implosion. Free radicals (e.g., •OH, •H) formed due to thermolysis of water molecules may react with the organic in the interfacial region or in the solution near the interface. Three primary pathways have been identified for compound degradation, including: (1) hydroxyl radical oxidation, (2) direct pyrolytic degradation, and (3) supercritical water reactions. In aqueous solution, water vapor present in the microbubble is homolytically split during bubble collapse to yield •H and •OH radicals, while chemical substrates present either within or near the gas-liquid interface of the collapsing microbubble are subject to direct attack by •OH. Volatile compounds such as benzene, toluene, ethyl benzene, xylene, and MTBE readily partition into the vapor of the growing cavitation microbubbles and then undergo direct pyrolysis during transient collapse.

Vapor stripping (air sparging) operations have as their goal to transfer the volatile contaminants from the liquid phase (i.e., groundwater) to the vapor phase (i.e., air). The combination of sonication and aeration/ozonation/hydrogen peroxide results in enhanced organic destruction and better liquid/vapor contact (i.e., better mixing) to facilitate the transfer of the degraded or partially degraded products into the gas phase. The addition of ozone (made through electric discharge of oxygen) into the liquid phase coupled with application of sonication considerably enhances the formation of hydroxyl radicals resulting in enhanced destruction of chlorinated organic compounds in solution. Hydroxyl radicals (•OH) are highly reactive non-specific reactants capable of oxidizing a wide variety of contaminants to carbon dioxide, hydrochloric acid (if chlorinated organics are present), and water. After mass transfer from the liquid phase into the gas phase, the water is pumped to a packer in the dual extraction well, resulting in the water (stripped of the volatile organic compounds (VOCs)) being returned into the subsurface, while the VOC-enriched vapor phase is passed through the second well and is sent to an above-ground vapor treatment system (that could involve capture on activated carbon/regeneration using ultrasonics, or thermal oxidation).

This research paper considered the following advanced oxidation treatment systems:

- Ultraviolet (UV) light;
- Hydrogen peroxide ($H_2O_2$)/UV light;
- Sonication;
- Sonication/UV light;
- Ozonation;
- Ozonation/$H_2O_2$; and
- Fenton's Reagent.

## Previous Studies with Evidence of Process Synergism

The photooxidation of *p*-xylene and *o*-nitrotoluene were studied by Ghaly [2] in a closed bench-scale unit using hydrogen peroxide or hydrogen peroxide/$Fe^{+2}$ as a photocatalyst. Results of the photo-Fenton process (UV/$H_2O_2$/$Fe^{+2}$) were compared with the UV/$H_2O_2$ process; the results indicated that the photo-Fenton process resulted in a higher efficiency degradation of the selected compounds and reduced energy consumption. The degradation reactions were 1$^{st}$ -order; the value of the reaction rate constant $k$ was ~5x higher for the photo-Fenton process than that in the case of the UV/$H_2O_2$ system. The quantity of both $H_2O_2$ and $Fe^{+2}$ must be properly selected the pH properly adjusted to optimize the process. For their particular system studied, the optimum reaction conditions were: 0.03 mole $H_2O_2$/L using the UV/$H_2O_2$ system and 0.01 mole $H_2O_2$/L, 1.0 mmole $Fe^{+2}$/L at pH ~3 for the photo-Fenton process.

The ozonation and peroxone oxidation of toluene in aqueous solutions was studied by Kuo and Chen [3]. They investigated the kinetics of the aqueous-phase oxidation or toluene by ozone and ozone-hydrogen peroxide mixtures at 25°C. The oxidation kinetics were 1$^{st}$-order with respect to ozone concentration; the reaction order in toluene varied with pH and the presence or absence of $H_2O_2$. The peroxone oxidation was half-order with respect to $H_2O_2$ in distilled water (initial pH of 5.4) and other solutions of higher pH. In acidic solutions with an initial pH $\leq 3$, the overall kinetics was 2$^{nd}$-order; the direct oxidation of toluene by ozone molecules was predominant in determining the slow rate of reaction. The reaction became very fast and enhanced with the addition of $H_2O_2$, if present, in alkaline solutions with an initial pH of 10 or higher. Under those conditions, the reaction was controlled by hydroxyl radical reactions and was independent of the toluene concentration.

## RESULTS AND DISCUSSION

### *Previous Studies Involving Combinations of Advanced Oxidation Studies*

**Figure 1** presents the results on the removal of atrazine from solution [4] using UV irradiation, ozonation, and combined UV/ozonation, while **Figure 2** presents analogous results for removal of carbofuran [5] using UV irradiation, Fenton's reagent, and combined UV/Fenton's treatment. **Figure 1** indicates that the removal of atrazine from solution is better than being additive, that much better performance is observed than predicted, indicating the process performed synergistically. Similar behavior is observed for the case of carbofuran. Beltran *et al.* [4] showed that the degradation efficiencies of atrazine by advanced oxidation treatment methods were in the following order: UV < $O_3$ < UV/$H_2O_2$ < UV/$O_3$. Similarly, Benitez *et al.* [5] observed that the degradation efficiencies of carbofuran were in the following order: UV < $O_3$ < Fenton's reagent < UV/$H_2O_2$ < UV/Fenton. Note that the integrated processes always had the best removal efficiencies of the target compounds.

**Figures 3 and 4** present the results for removal of carbon tetrachloride ($CCl_4$) and trichloroethane (TCA) from solution using combined sonication and air stripping operations [6, 7]; the prediction from the additivity of the 1$^{st}$-order rate constants is shown by the curve in the figure; the actual observed data is shown as the individual data points. Once again, the combination of the two treatment processes (sonication and air stripping) is seen to operate in a synergistic fashion.

**Table 1** compares the 1$^{st}$-order rate constants using sonication, air sparging, and their combination for removal of benzene, toluene, ethyl benzene, and *o*-xylene from solution [1]. Once again, the combination of these treatment processes is considerably better than the individual processes, and operates in a synergistic fashion, as evidenced by the observed 1$^{st}$-order rate constants being higher than the predicted values.

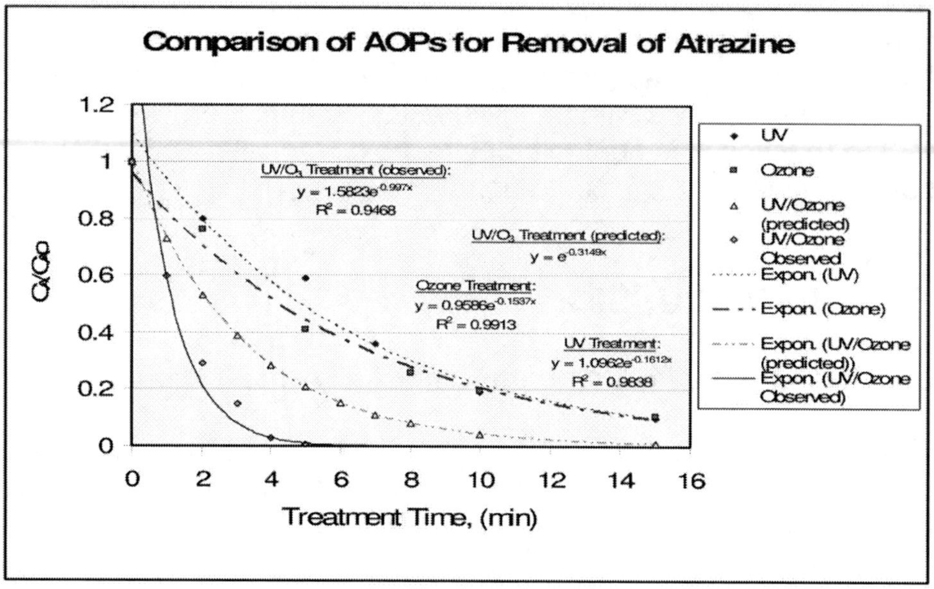


**Figure 1.** Comparison of AOPs for removal of atrazine from solution, adapted from [2].

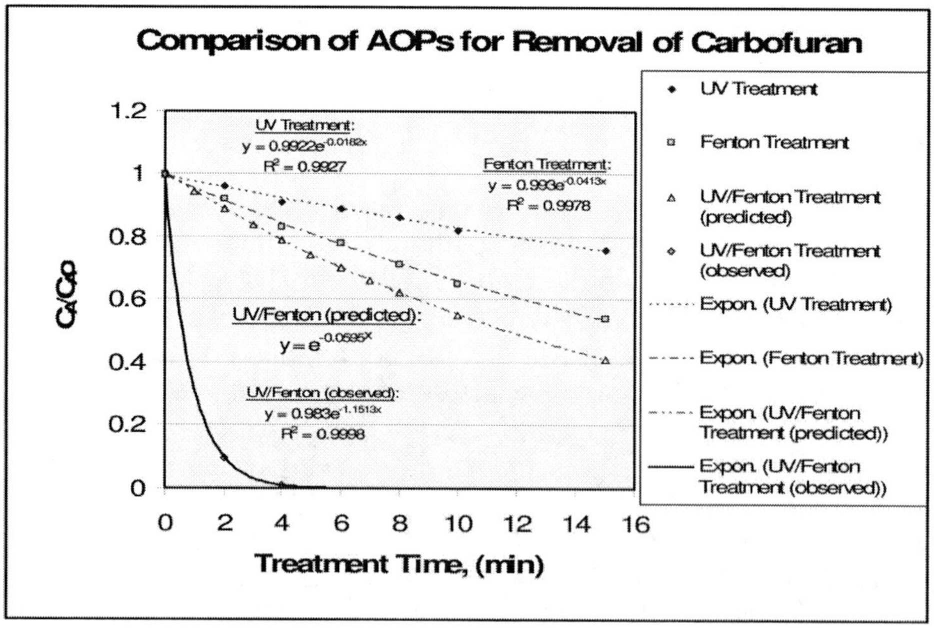


**Figure 2.** Comparison of AOPs for removal of carbofuran from solution, adapted from [3].

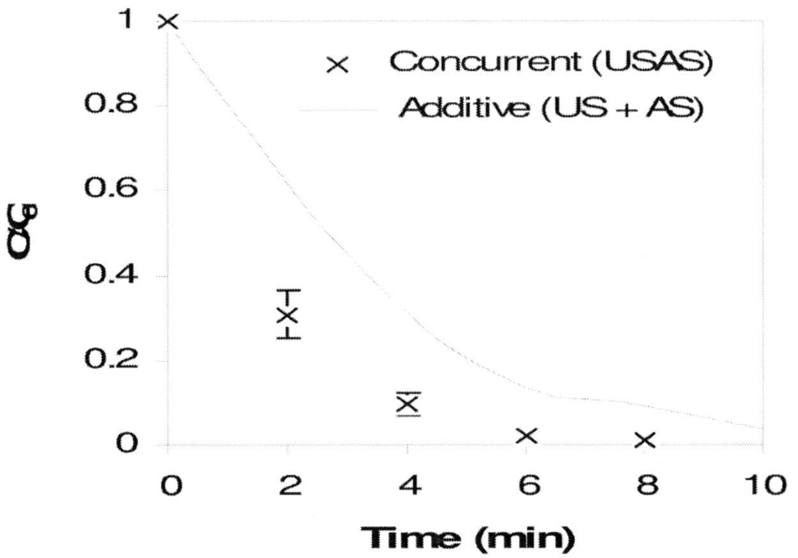


**Figure 3.** Removal of $CCl_4$ from artificial groundwater by combined ultrasound and air stripping: Ultrasonic power intensity = 35 W/cm$^2$ and air flow rate = 500-mL/min.

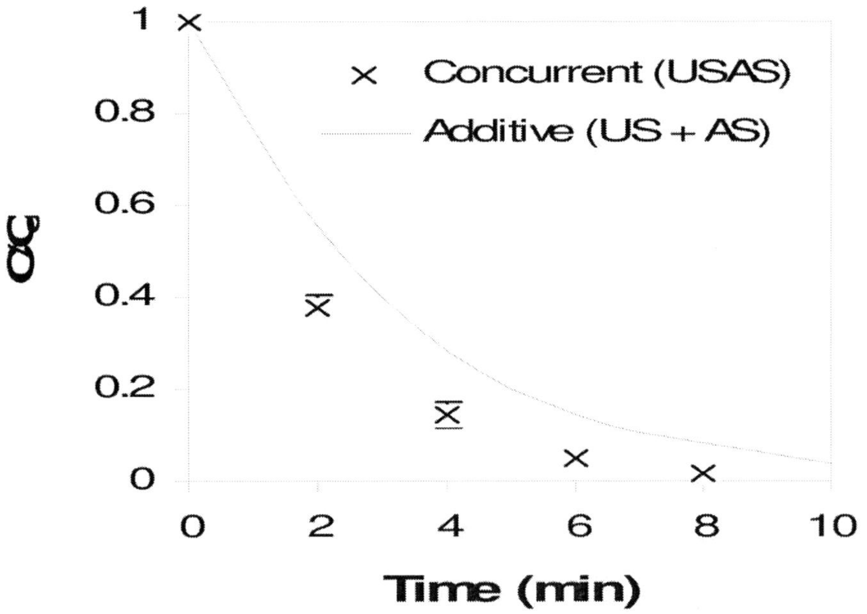


**Figure 4.** Removal of TCA from artificial groundwater by combined ultrasound and air stripping: Ultrasonic power intensity = 35 W/cm$^2$ and air flow rate = 500-mL/min.

**Table 1.** Summary of 1st-order rate constants using various treatment processes for removal of benzene, toluene, ethyl benzene, and *o*-xylene from solution [1].

| Contaminant | Treatment Process | 1st-Order Rate Constant, ($min^{-1}$) |
|---|---|---|
| Benzene | Sonication alone | 0.0608 |
| | Air sparging alone | 0.1347 |
| | Combined sonication + air sparging (predicted) | 0.1955 |
| | Combined sonication + air sparging (observed) | 0.2166 |
| Toluene | Sonication alone | 0.0635 |
| | Air sparging alone | 0.1381 |
| | Combined sonication + air sparging (predicted) | 0.2016 |
| | Combined sonication + air sparging (observed) | 0.2443 |
| Ethyl benzene | Sonication alone | 0.0702 |
| | Air sparging alone | 0.1655 |
| | Combined sonication + air sparging (predicted) | 0.2357 |
| | Combined sonication + air sparging (observed) | 0.3684 |
| *o*-Xylene | Sonication alone | 0.0497 |
| | Air sparging alone | 0.1385 |
| | Combined sonication + air sparging (predicted) | 0.1882 |
| | Combined sonication + air sparging (observed) | 0.2132 |

## Summary and Conclusions

For the research investigations studied, integrated processes generally perform in a synergistic manner, rather than giving the performance of the individual processes simply being additive. For example, a synergistic effect is observed in the photo-Fenton process which was attributed to the generation of hydroxyl radicals (•OH) in several reactions and to the catalytic behavior of the Fenton's reaction in the presence of UV radiation. Thus, for the oxidant concentrations employed in the Benitez study [5], the efficiency of the photo-Fenton process was even higher than that of the conventional ozonation and the $UV/O_3$ system. As shown in the analyses presented in this paper, the rate constant of the 1st-order combined treatment system was 3.167x and 19.35x greater than predicted rate constant (the sum of the individual unit processes rate constants) for the removal of atrazine and carbofuran, respectively. For the removal of $CCl_4$ and TCA [6, 7], the observed rate constant was 1.777x and 1.653x times greater than the predicted value (the sum of the 1st-order rate constants for sonication and air stripping), respectively. Similarly, for treatment of benzene, toluene, ethyl benzene, and *o*-xylene, the observed rate constant [1] was 1.11x, 1.21x, 1.56x, and 1.13x times greater, respectively, than the predicted value (the sum of the 1st-order rate constants for sonication and air stripping). Sonication by itself is not a highly effective process, while combined processes (e.g., sonication + air sparging, sonication + $H_2O_2$, sonication + UV, etc.) are synergistic in nature, providing much greater treatment effectiveness than the efficiencies predicted by the sum of the 1st-order rate constants. The improvement in rate constants resulting from integrated processes can be an order-of-magnitude or more high than those of single processes alone (such as sonication or air sparging). These combined, integrated unit operation processes operated in a synergistic fashion.

## References


[1] Peters, R.W., Gauthier, J.J., and Mohammad, J., (2006). "Final Report: Use of Sonication/Acoustic Cavitation with Advanced Oxidants to Treat Petroleum Hydrocarbons-Contaminated Surface Waters and Groundwaters", Report submitted to the Alabama Water Resources Research Institute.

[2] Ghaly, M.Y., (2005). "Destruction of the Environmental Organic Pollutant through Activated Hydrogen Peroxide Oxidation", *Afinidad, 62*(518): 314-320.

[3] Kuo, C.-H., and Chen, S.-M., (1996). "Ozonation and Peroxone Oxidation of Toluene in Aqueous Solutions", *Indus. Eng. Chem. Res.*, *35*(11): 3973-3983.

[4] Beltran, F.J., Garcia-Araya, J.F., and Acedo, B., (1994). "Advanced Oxidation of Atrazine in Water – II. Ozonation Combined with Ultraviolet Radiation," *Water Research*, *28*(10): 2165 – 2174.

[5] Benitez, F.J., Acero, J.L., and Real, F.J., (2002). "Degradation of Carbofuran by Using Ozone, UV Radiation, and Advanced Oxidation Processes", *J. Hazard. Mater.*, *89*(1): 51-65.

[6] Peters, R.W., and Sharma, M.P., (2006). "A Critical Evaluation of Integrated Advanced Oxidation Processes: Assessment of Process Synergism", Paper presented at the 2006 AIChE Meeting, San Francisco, CA, (November 12-17).

[7] Ayyildiz, O., (2003). "Ultrasonic and Air Stripping Removal of $CCl_4$ and TCA from Water: Investigating Synergistic Effects", Ph.D. Dissertation, Illinois Institute of Technology, Chicago, IL, (July).

# Filtration of Modified Monosodium Titanate Slurries

**Michael Poirier**, Savannah River National Laboratory, Savannah River Site, Aiken, SC 29808 and Samuel D. Fink, Environment & Chemical Process Technology, Savannah River National Laboratory, Aiken, SC 29808.

Filtration of Modified Monosodium Titanate Slurries

# Novel Separation Process of Gaseous Mixture of $SO_2$ and $O_2$ with Ionic Liquid for Hydrogen Production in Thermochemical Sulfur-Iodine Water Splitting Cycle

Chang Soo Kim[1], Ki Yong Lee[2], Kwang Ho Song[2], Gyeong Taek Gong[1], Kye Sang Yoo[1], Honggon Kim[1], Byoung Gwon Lee[1], Byoung Sung Ahn[1], and Kwang Deog Jung[1]
[1]Clena Energy Research Center, Energy and Environment Technology Division, Korea Institute of Science and Technology, Hawolgokdong 39-1, Seongbukgu, Seoul 139-791, South Korea
[2] Department of Chemical and Biological Engineering, Korea University, Anamdong, Seongbukgu, Seoul 136-701, South Korea

Sulfur-Iodine cycle is the most promising thermochemical cycle for water splitting to produce hydrogen which can replace the fossil fuels in the future. As a sub-cycle in the thermochemical Sulfur-Iodine water splitting cycle, sulfuric acid ($H_2SO_4$) decomposes into oxygen ($O_2$) and sulfur dioxide ($SO_2$) which should be separated for the recycle of $SO_2$ into the sulfuric acid generation reaction (Bunsen Reaction).

Conventional gas separation processes are Distillation, Pressure Swing Adsorption (PSA), Temperature Swing Adsorption (TSA), and solvent absorption. Distillation, PSA and TSA are basically energy intensive processes to control the pressure and temperature in a large system. Absorption process is not attractive for the high temperature process because of the volatile property of the solvent. A novel separation process needs to be proposed to compensate the demerits of conventional gas separation methods for the high temperature-gaseous $SO_2/O_2$ mixture.

Ionic Liquid is not volatile even at high temperature, which can improve absorption process utilizing organic solvents by eliminating the issue of volatility of the solvents. In addition, a continuous process can be realized by recycling non-volatile ionic liquid at high temperature. The cost of gas separation process can be reduced by avoiding Distillation, PSA and TSA by applying absorption system at high temperature without absorbent loss by ionic liquid absorbent compared to the solvent absorbent which much volatile than ionic liquids.

For FGD (Flue Gas Desulfurization), the possibility of the use of ionic liquids as absorbents are investigated [1-3]. Most of the reports are related to the absorption mechanism whether it is chemical or physical. Recently, researchers have announced that physical absorption of $SO_2$ in ILs is verified by the IR spectra and reversible absorption which can be useful for the desulfurization process [3].

In this study, absorption and desorption process of $SO_2$ by ionic liquid which is useful for the recycle of $SO_2$ into sulfuric acid generation reaction after sulfuric acid decomposition in the thermochemical Sulfur-Iodine cycle is investigated. At first, the operability as an absorbent for the $SO_2$ absorption and desorption at high temperature without the volatilization of absorbents which is not suitable for the recycle of absorbent-free $SO_2$ after the absorption process. In Fig. 1, the temperature range of operability is determined by TGA and DTA analysis. Most of ionic liquids investigated are applicable at high temperature desorption without volatility around 300°C except [BMIm] Cl, and [BMIm] OAc which show the decomposition of ionic liquids.

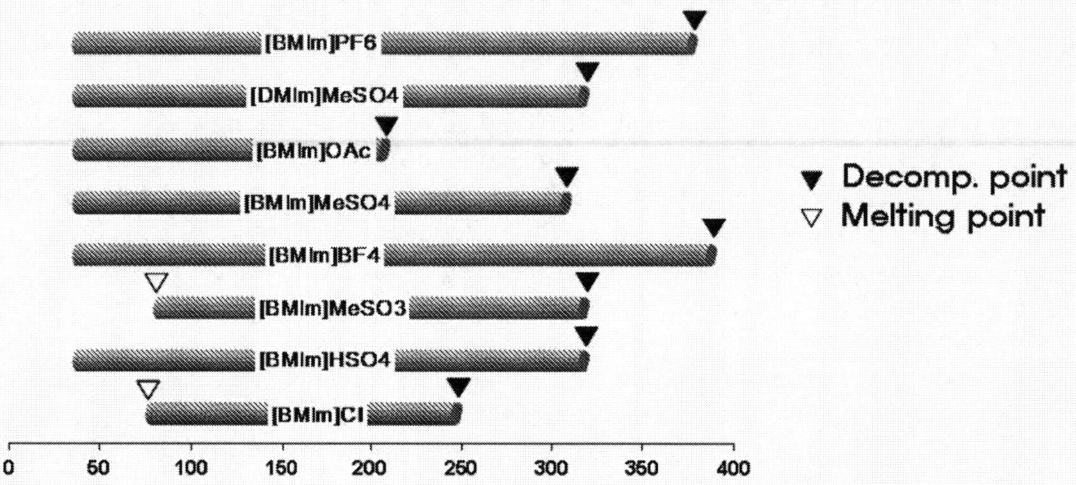


**Figure 1** Operable temperature range of each ionic liquid without decomposition measured by TGA and DTA.

To evaluate the capability of $SO_2$ absorption, each ionic liquid is located in the absorption tube in Fig. 2 and gaseous $SO_2$ is bubbled into the ionic liquid. During the bubbling, the weight of the system is measured and converted into the absorbed $SO_2$ amount at each temperature controlled by the heater.

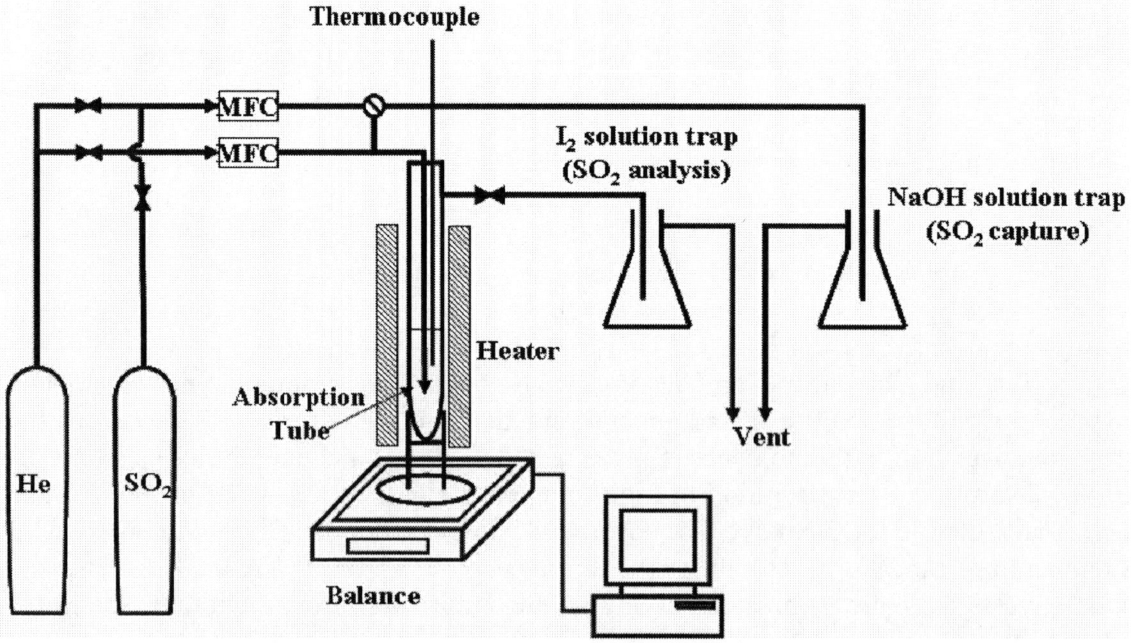


**Figure 2** Schematic of $SO_2$ absorption system

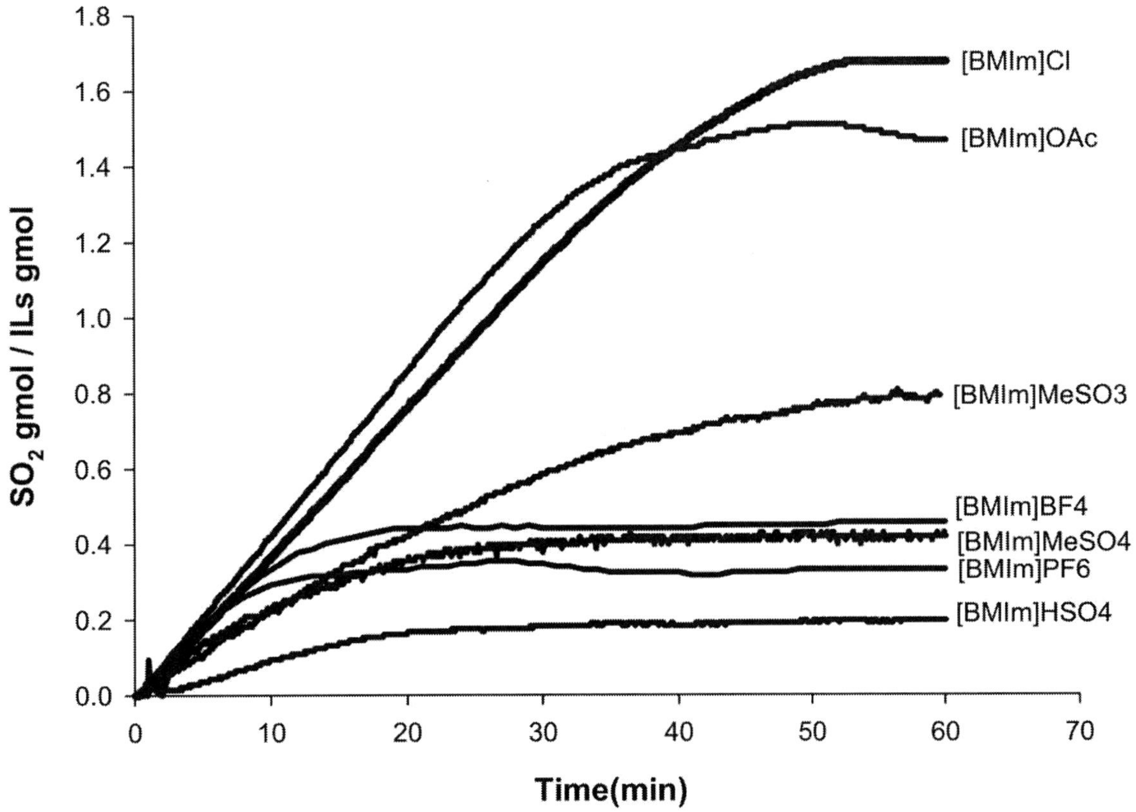


**Figure 3** $SO_2$ absorption curve of each ionic liquid at 50°C with flow rate of $SO_2$ 20 $cm^3$/min and He 10 $cm^3$/min at 1 atm.

Saturated amounts of absorbed $SO_2$ by ionic liquids at 50°C are presented in Fig.3. The effect of anions for the $SO_2$ absorption capability is shown in the order of Cl, OAc, MeSO3, BF4, MeSO4, PF6, and HSO4 when they are combined with [BMIm] cation. [BMIm]Cl has the largest amount of $SO_2$ absorbed which can be the most promising absorbent; however, from the point of operability at high temperature which includes desorption process, [BMIm]Cl is vulnerable to high temperature around 250°C based on the TGA/DTA analysis in Fig. 1. As the second largest $SO_2$ absorbing ionic liquid, [BMIm]OAc is not stable at the temperature around 200°C from Fig. 1, either.

The effect of temperature for $SO_2$ absorption by ionic liquids at 1 atm is presented in Fig. 4 with respect to ionic liquids. The amount of absorbed $SO_2$ decreases as temperature increases for all ionic liquids experimented. At low temperature the difference of the absorbed $SO_2$ between each ionic liquid is bigger than that of at high temperature.

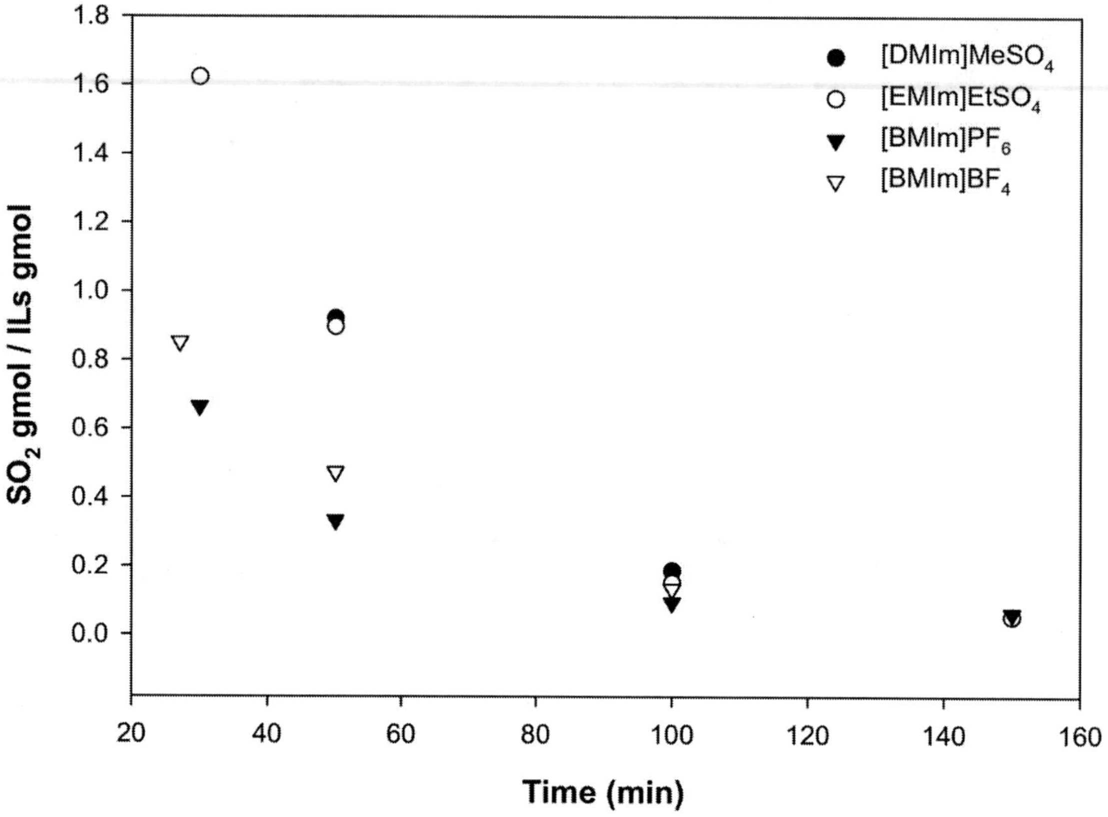


**Figure 4** Absorbed $SO_2$ ratio in ILs as a function of time at 1 atm

($SO_2$ : 20 $cm^3$/min, He : 10 $cm^3$/min)

The effect of partial pressure of $SO_2$ to the absorbed $SO_2$ amount is presented in Fig. 5. The partial pressure of $SO_2$ is controlled by changing the ratio between inert He and $SO_2$. The amount of absorbed $SO_2$ increases as partial pressure of $SO_2$ increases for [DMIm] MeSO4. The trends of $SO_2$ absorption depending on the temperature which is mentioned earlier are also recognized in Fig. 5.

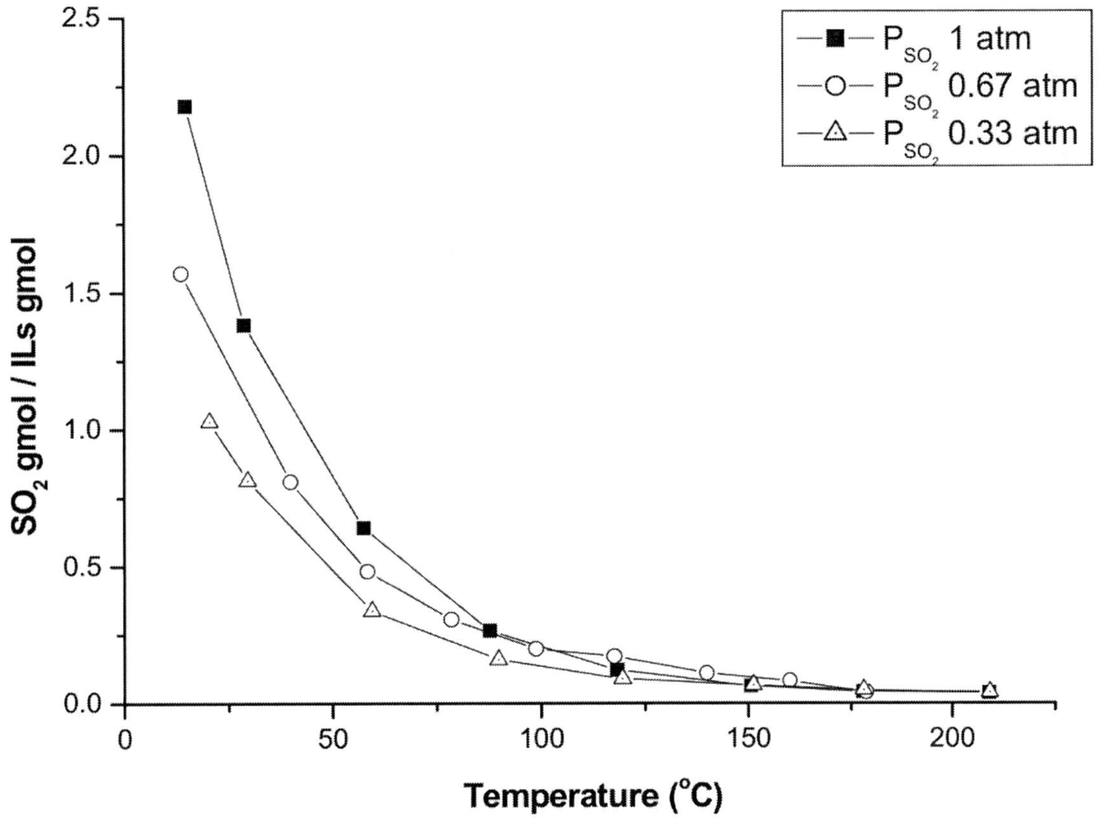


**Figure 5** Absorbed $SO_2$ ratio in [DMIm] MeSO4 as a function of partial pressure

Until now, the solubility of $SO_2$ in the ionic liquids is investigated. However, in the sulfur cycle, oxygen exists together with $SO_2$ as the products of sulfuric acid decomposition after $H_2O$ removal by condensation. The solubility of oxygen in the ionic liquid, which can be expected as negligible because of its non-polarity of the molecule, is measured with the same apparatus described in Fig. 2 by changing the feeding gas into oxygen. Solubility of oxygen and $SO_2$ is compared in Table 1 where [EMIm] EtSO4 is used as an absorbent at 1 atm 50°C. The amount of absorbed oxygen is much less than that of $SO_2$ by the order difference.

**Table 1** Comparison of the absorbed amount of $SO_2$ and $O_2$ by [EMIm]EtSO4 at 1 atm

| Temperature (°C) | $SO_2$ mol/ IL mol | $O_2$ mol/IL mole |
|---|---|---|
| 50 | 0.90 | 0.08 |

The absorption mechanism of $SO_2$ in the ILs is known as physical absorption by many researchers [2,3] these days, where chemical absorption is also reported with some ionic liquids [1]. Physical absorption property of ionic liquids is necessary for the recycle process of the absorbed pure $SO_2$ into sulfuric acid generation reaction. By the FT-IR spectra, the physical absorption is verified with the fact that no spectra peak position change of ionic liquids before and after $SO_2$ absorption except the generated $SO_2$ peak in the spectrum. The effect of anion in the interaction between $SO_2$ and ionic liquids during absorption process is analyzed by FT-IR spectra in Fig. 6. Peak shift, which is presented in the rectangle, is recognized by comparing $SO_2$ absorbed [BMIm] PF6 and [BMIm] BF4 spectrum. The amount of interaction which results in the absorption amount difference in Fig. 3 may not analyzed quantitatively; however, it can be verified qualitatively because only anion is the difference between two spectrum conditions.

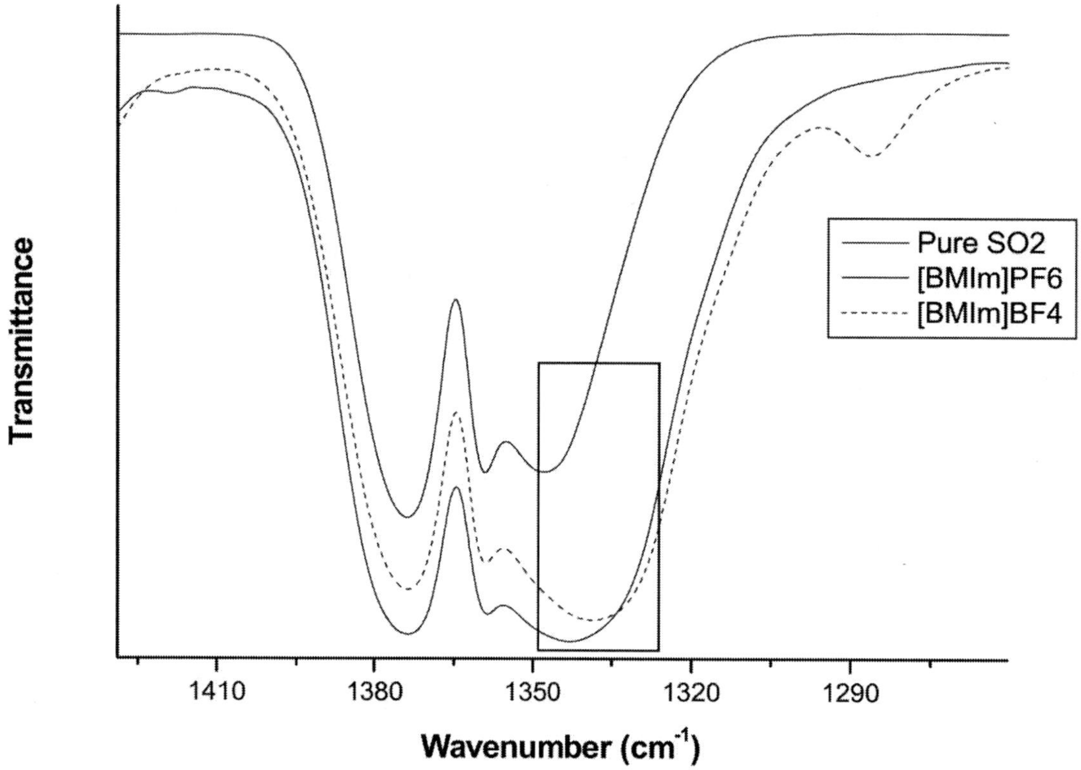


Figure 6 IR spectra of pure $SO_2$, $SO_2$ absorbed [BMIm]PF6, and [BMIm]BF4

Based on the properties of $SO_2$ absorbing ionic liquid mentioned, continuous $SO_2$ separation process of the gaseous $SO_2/O_2$ mixture in the thermochemical sulfur-iodine cycle can be realized after acquiring more properties of ionic liquid which is useful for the process design in the near future.

**References**


1. Weize Wu, Buxing Han, Haixiang Gao, Zhimin Liu, Tao Jiang, and Jun Huang. Desulfurization of Flue Gas: SO2 Absorption by an Ionic Liquid. *Angewandte Chemie* . 2004; 43: 2415

2. Jessica L. Anderson, JaNeille K. Dixon, Edward J. Maginn, and Joan F. Brenecke. Measurement of SO2 Solubility in Ionic Liquids. *J. Phys. Chem. Lett. B*. 2006; 110: 15059
3. Jun Huang, Anders Riisager, Peter Wasserscheid, and Rasmus Fehrmann. Reversible Physical Absorption of SO2 by Ionic Liquids. *Chem. Comm.* 2006; 4027

# MELCOR-H2 Benchmarking of the SNL Transient Sulfuric Acid Decomposition Experiments

*Sal B. Rodriguez[1], David Louie[2], Randall O. Gauntt[1], Fred Gelbard[1], Paul Pickard[1], Randy Cole[1], Katherine McFadden[1], Tom Drennen[1], Billy Martin[1], Louis Archuleta[2], Shripad T. Revankar[3], Karen Vierow[4], Mohamed El-Genk[5], and Jean Michel Tournier[5]*
*P.O. Box 5800, MS 0748, Albuquerque, NM 87185-0748, sbrodri@sandia.gov*
*Sandia National Laboratories[1]*
*OMICRON Safety and Risk[2]*
*Purdue University[3]*
*University of Texas A & M[4]*
*University of New Mexico[5]*

## Abstract

MELCOR is a world-renowned nuclear reactor safety analysis code that is used to simulate both light water and gas-cooled reactors. MELCOR-H2 is an extension of MELCOR that can model detailed nuclear reactors that are fully coupled with modular secondary-system components and the sulfur iodine (SI) thermochemical cycle for the generation of hydrogen and electricity. The models are applicable to both steady state and transient calculations. Previous work has shown that the hydrogen generation rate calculated by MELCOR-H2 for the SI cycle was within the expected theoretical yield, thus providing a macroscopic confirmation that MELCOR-H2's computational approach is reasonable. However, in order to better quantify its adequacy, benchmarking of the code with experimental data is required.

Sulfuric acid decomposition experiments were conducted during late 2006 at Sandia National Laboratories, and MELCOR-H2 was used to simulate them. We developed an input deck based on the experiment's geometry, as well as the initial and boundary conditions, and then proceeded to compare the experimental acid conversion efficiency and $SO_2$ production data with the code output. The comparison showed that the simulation output was typically within less than 10% of experimental data, and that key experimental data trends such as acid conversion efficiency, molar acid flow rate, and solution mole % were computed adequately by the MELCOR-H2.

[a]Sandia is a multiprogram laboratory operated by Sandia Corporation, a Lockheed Martin Company, for the United States Department of Energy's National Nuclear Security Administration under Contract DE-AC04-94AL85000.

## Introduction

Large-scale production of hydrogen through nuclear/thermochemical plants is the subject of much recent interest. This is particularly so of the sulfur iodine (SI) cycle and the secondary systems interconnected to it [1-3]. Other important thermochemical cycles such as the Westinghouse Hybrid Sulfur (HyS) and adiabatic UT-3 cycles are being considered [1, 3]. It is significant that both the SI and Westinghouse cycles share the same sulfur chemistry, so advances in one naturally lead to advances in the other. A literature search shows many recent efforts to model SI and HyS chemistry that is coupled to secondary systems, but these employ primarily steady-state models, and/or are not coupled to a fully-detailed nuclear plant [1-4]. Fully-coupled, dynamic modeling is important, as current research shows that there is a complex, dynamic interdependent behavior of the subsystems, and this affects hydrogen and electrical output, safety, and overall plant behavior.

MELCOR has been used to adequately simulate numerous types of nuclear reactors, including high-temperature gas-cooled reactors [5-7]. MELCOR-H2 is an extended version of MELCOR designed to fully couple detailed models of nuclear reactors with modular secondary-system components and thermochemical cycles for the generation of hydrogen and electricity [8-13].

Whereas previous MELCOR-H2 work has shown that it can yield hydrogen in amounts corresponding to expected theoretical yields [14, 15], the next step is to show that the MELCOR-H2 SI transient chemistry models can adequately simulate experimental data. To that effect, we simulated the sulfur decomposition experiments that were conducted at Sandia National Laboratories (SNL) recently [16, 17]. This paper briefly discusses the experiments, and then continues with a description of the MELCOR-H2 models, how the simulations were conducted, and the results of the simulations.

## Brief Experiment Description

SNL is currently advancing the production of hydrogen for the SI cycle by investigating the decomposition of sulfuric acid. In the first series of experiments, a half scale decomposer experiment was fabricated. It consists of a bayonet that is made up of three vertical concentric silicon carbide (SiC) tubes, as shown schematically in Figs. 1 and 2. The cross sectional dimensions are shown; the length of the tube was 27". Ceramic materials such as SiC were used to construct the experiment because of its resistance to corrosion by acid. However, the tubing is difficult to fabricate and brittle. This makes it somewhat difficult to add thermocouples and other instruments in order to acquire more detailed data for numerical analysis.

The acid solution entered at the bottom of the bayonet in the Teflon manifold and flowed up along the outer annulus of the tubes. The inlet temperature was set at 30 °C. While the liquid acid solution was flowing up, the recuperation of the heat from the downward flow of the outgoing products and un-reacted acid solution and the external electric heater were sufficient to bring the upward flow of the acid solution to boil and to dissociate the sulfuric acid into sulfur trioxide and water vapor. The temperature of the incoming solution reached about 850 °C, and thus became a superheated acid vapor, before entering the catalyst bed.

The catalyst was about a third of the length of the entire assembly, was made of a quartz baffle that held up the rows of the catalyst platinum pellets, and had a porosity of 50%. The sulfur trioxide was decomposed into sulfur dioxide and oxygen gases as it flowed through the catalyst. Any un-reacted sulfur trioxide combined with water to form back the acid while the flow was cooled down by the recuperation process in the inner annulus of the tubes. The expected exit temperature of the products and un-reacted acid solution was about 180 °C.

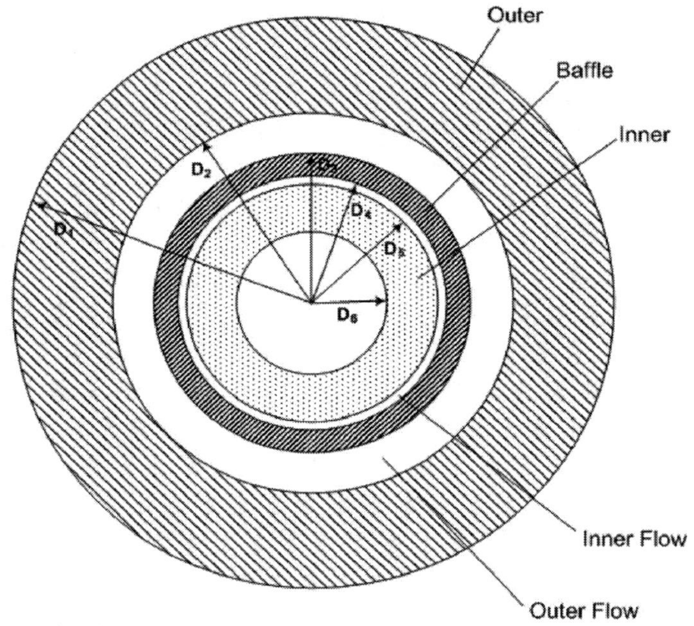


Outer: OD 38.1 mm ($D_1$)
ID 25.4 mm ($D_2$)

Baffle: OD 20 mm ($D_3$)
ID 17 mm ($D_4$)

Inner: OD 15.875 mm ($D_5$)
ID 9.525 mm ($D_6$)

Outer Flow Annulus $D_{o1} = D_2$
$D_{i1} = D_3$

Inner Flow Annulus $D_{o2} = D_4$
$D_{i2} = D_5$

**Fig. 1. Dimensions for the Bayonet Section.**

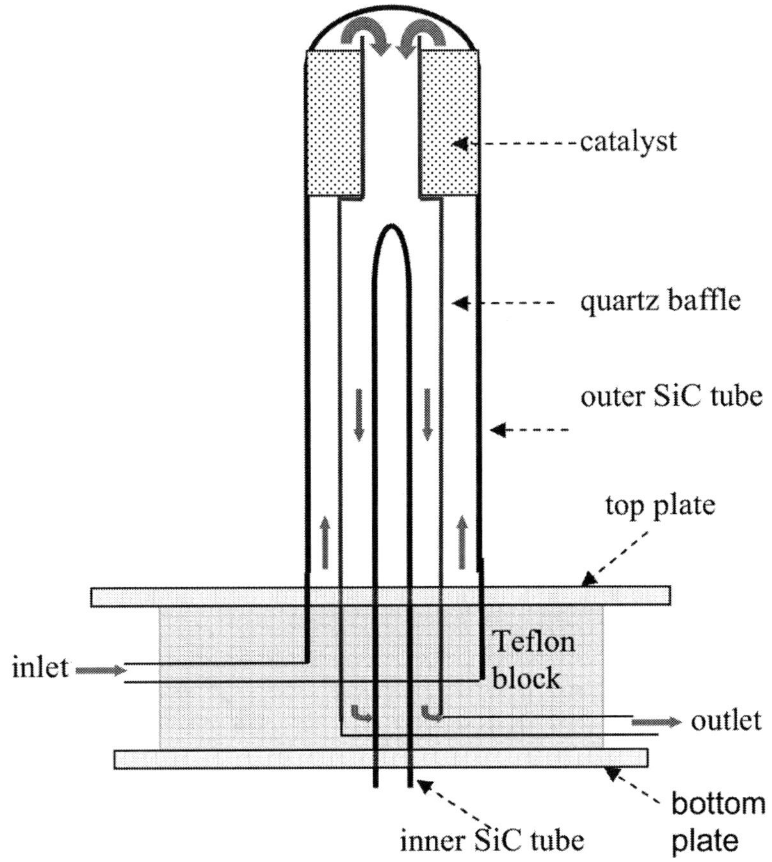


**Fig. 2. Bayonet Schematic.**

Based on the dimension of the tubes in Fig. 1, the upward flow volume of the bayonet was about 123 $cm^3$, and the downward flow volume was about 43.6 $cm^3$. Thus, the total flow volume in the bayonet was about 166.6 $cm^3$.

A series of test runs for the half scale experiment were conducted at 850 °C, ambient pressure, with inlet acid solutions of 19, 38, 42, and 53 mole %. Note that the data for 53 mole % were not simulated, as they appeared to be atypical from the 19, 38, and 42 mole % experiments; more investigation is needed to determine what variables may have contributed to such behavior. A wide range of solution molar rates were considered, and can be found in references [16, 17]. A more thorough description of the experiment is outside the scope of this paper; however, interested readers are referred to references [16, 17].

## Modeling

The chemical reaction for the decomposition of the sulfuric acid is given below:

$$H_2SO_4 \Rightarrow SO_2 + \tfrac{1}{2}O_2 + H_2O \qquad (1)$$

Equation (1) considers the stoichiometric reaction because 1 mole of $H_2SO_4$ will yield 1 mole of $SO_2$, a half mole of $O_2$, and 1 mole of $H_2O$. There are two principle reactions that contribute to Equation (1):

$$H_2SO_4 \Leftrightarrow SO_3 + H_2O \tag{2}$$

The dissociation of $H_2SO_4$ as shown in Equation (2) takes place above 337 °C. If the temperature decreased below this value, the formation of $H_2SO_4$ from $SO_3$ and $H_2O$ will take place. In most cases, this reaction occurs instantaneously. If the temperature continued to rise, and with the help of the catalyst (in a catalyst bed), the decomposition of $SO_3$ will occur:

$$SO_3 \Rightarrow SO_2 + \tfrac{1}{2}O_2 \tag{3}$$

Once the $SO_3$ decomposes as shown in Equation (3), the reaction is usually irreversible. The desired reaction temperature is about 850 °C, which would yield a high conversion efficiency. Thus, the determining reaction for the decomposition of the sulfuric acid depends on Equation (3). The rate constant for Equation (3) is usually given in the Arrhenius form as given below. MELCOR-H2 version 1 models the decomposition of $H_2SO_4$, using the reaction rate constant, which is given as follows,

$$k = A \cdot e^{\frac{-E}{R \cdot T}} \tag{4}$$

where A is a pre-exponential or frequency factor ($s^{-1}$), E is the activation energy (J/mole), R is the gas constant (8.314 J/mole-K), and T is the reaction temperature (K). E is given as $7.31\times10^4$ J/mole-K. The value of A is taken to be $6.8\times10^4$ $s^{-1}$, which is based on the decomposition rate of $SO_3$ into $SO_2$ and $O_2$, and with the assumption of instantaneous conversion of $H_2SO_4$ to $H_2O$ and $SO_3$ [18]. Thus the efficiency for this conversion is simply given by the following equation:

$$\eta = 100\% \times (1 - e^{-k \cdot t_r}) \tag{5}$$

where $\eta$ is the efficiency (%), k is given in Equation (4), and $t_r$ is the residence time (s) of the acid that contacts the catalyst bed where the decomposition is promoted. The equations show that the reaction rate is strongly dependent on the reaction temperature—the higher the reaction temperature, the higher the conversion efficiency.

To simulate the half scale bayonet experiment, MELCOR-H2 was modified to include a model for a flow system of the chemical species, rather than using single point "chemistry nodes". The principal equations for modeling this experiment are briefly described below.

For a given inlet acid solution flow rate, $\dot{M}_{inlet}$ with $f_1$ (the mole fraction of $H_2SO_4$), the $H_2O$ and acid inlet flow rate are given, respectively, as:

$$\dot{M}_{H_2O} = \dot{M}_{inlet}(1\text{-}f_1) \tag{6}$$

$$\dot{M}_{acid} = \dot{M}_{inlet} \times f_1 \tag{7}$$

Let $M_{acid,r}$ be the number of moles of acid actually decomposed, and according to Equation (1), the number of moles for the products is given as

$$M_{SO_2} = M_{acid,r} \cdot 1$$
$$M_{O_2} = M_{acid,r} \cdot 0.5 \tag{8}$$
$$M_{H_2O} = M_{acid,r} \cdot 1$$

Thus, for each mole of $H_2SO_4$ that is decomposed, 2½ moles of products are created. The conversion efficiency (%), is defined as

$$\eta_{acid} = \frac{M_{acid,r}}{f_1 \times \dot{M}_{inlet} \times t_r} \times 100\% \tag{9}$$

Note that both Equations (5) and (9) calculate the conversion efficiency. When the residence time $t_r$ is known, the amount of acid decomposed can be calculated easily for a given temperature.

Once the amount of products is calculated, the outlet flow rate of the products, including any un-reacted acid is given by the following relations:

$$\dot{M}_{acid,unreacted} = \dot{M}_{acid} \times (1-\eta_{acid}) \tag{10}$$

$$\dot{M}_{SO_2} = \dot{M}_{acid} \times \eta_{acid} \tag{11}$$

$$\dot{M}_{O_2} = \dot{M}_{acid} \times \eta_{acid} \times \frac{1}{2} \tag{12}$$

$$\dot{M}_{H_2O}^{(outlet)} = \dot{M}_{inlet}(1-f_1) + \dot{M}_{acid} \times \eta_{acid} \tag{13}$$

Note that Equation (13) includes both inlet $H_2O$ and the amount of $H_2O$ produced. Thus the total outlet flow rate is the sum of Equations (10) through (13).

In terms of energy, the upward flow of the inlet acid solution will heat up from 30 °C to the reaction temperature (maximum of 850 °C). Both the sensible heat for the acid and water were calculated, starting from their liquid state to the final gas state. At the catalyst location, an additional energy term is estimated for the reaction (dissociation energy). As the reaction products, the un-reacted acid, and water vapor flow downward to the exit, the recuperation process will transfer heat from the incoming fluid as shown in Figure 2. This transfer is also captured in MELCOR-H2.

Note that the primary goal for the SNL experiment was to advance the technology for producing larger quantities of hydrogen, as opposed to providing a fully-instrumented experiment for the purpose of code validation. As such, then, the full nature of the fluid entering the catalyst was not recorded experimentally; in particular, it was unclear if it was completely gaseous, or if it consisted of two phases (liquid and gas). This information was required in order to better estimate the fluid $t_r$ as it traveled through the catalyst. We decided to use the Ergun equation as a first-cut approximation [19],

$$\left[\frac{(P_0-P_L)\rho}{G_0^2}\right]\left(\frac{D_p}{L}\right)\left(\frac{\varepsilon^3}{1-\varepsilon}\right)=150\left(\frac{1-\varepsilon}{D_pG_0/\mu}\right)+\frac{7}{4}, \tag{14}$$

where P is the pressure at a given position, ρ is the density, $D_p$ is the pellet diameter, $G_0$ is the mass flux, L is the catalyst length, μ is the fluid viscosity, and ε is the catalyst porosity. By employing the Ergun equation, we obtained an approximate value for $t_r$, which we then corroborated by running a MELCOR-H2 simulation and then backtracking $t_r$. The two were found to be consistent. Then, once $t_r$ was estimated for a given run, $t_r$ for the remaining experimental runs was prorated as follows,

$$t_{r,new}=t_{r,max}\left(\frac{\dot{M}_{max}}{\dot{M}_{new}}\right) \tag{15}$$

where the subscript "max" indicates that the molar flow rate was the maximum used in the experiment and $t_r$ corresponded to that molar flow rate. This is a first-cut approximation to $t_r$, but sufficient, given the amount of experimental data that was available. In short, the simulation of each experimental run employed a unique $t_r$ because each was conducted at a different molar flow rate.

For each experimental run, the measured inlet acid solution in mole % of $H_2SO_4$ was included in the input deck. Beginning with MELCOR-H2 version 1, the code permits the modeling of the flow system, such as the experiment described above. Thus, the required input for two chemical species was needed: $H_2SO_4$, and $H_2O$. For a given acid inlet flow rate ($\dot{M}_{H_2SO_4}$), the corresponding water component flow rate ($\dot{M}_{H_2O}$) in the acid solution was computed according to the following equation:

$$\dot{M}_{H_2O}=\frac{\dot{M}_{H_2SO_4}\times(1-f_1)}{f_1} \tag{16}$$

where $f_1$ is the mole fraction of the acid solution. For example, suppose there is an acid inlet flow rate of 3.1 mole/hr ($8.61\times10^{-4}$ mole/s), than a 38 mole% solution has 5.06 mole/hr of water ($1.40\times10^{-3}$ mole/s). These flow rates are included as input, along with the temperatures of the experiment. The inlet temperature is taken to be 30 °C. The reaction temperature was input for the desired temperature of 850 °C, although the code recalculates the reaction temperature based on the heat transfer characteristics and flow rates. The outlet temperature was taken to be at 180 °C. The volume of the bayonet was not needed, since the rate constant for this reaction was independent of volume. If desired, this volume may be used as part of an alternative estimation for $t_r$. Additional inputs are required to model the outlet flow conditions, beside the outlet temperature given above. Because this is a flow system with no accumulation, any material that goes in should come out from the bayonet. Therefore, the outlet flow rates of the reactants and products were modeled according to the given inlet condition, and the assumption of 100% reaction completion: that is, for 1 mole of $H_2SO_4$ decomposed, there should be 1 mole of $SO_2$, 1 mole of $H_2O$, and 0.5 mole of $O_2$. Then, for example, a stoichiometric balance for 3.1 mole/hr of $H_2SO_4$ flowing in should result in the production of 3.1 mole/hr of $SO_2$, 3.1 mole/hr of $H_2O$, and 1.55 mole/hr for $O_2$. For the total $H_2O$ flow out (sum of what is flowing in and what is produced), the molar flow rate is 8.16 mole/hr (3.1 mole/hr production + 5.06 mole/hr inlet).

Whereas MELCOR-H2 can simulate both sulfuric and hydriodic (HI) acid decomposition, as well as the Bunsen reaction, only the sulfuric acid section was modeled, as the SNL experiments did not include the latter two. In the future, we will simulate those sections, as experimental data becomes available.

The simulations were run for 100 transient seconds, by which time the calculation had reached an asymptotic state. The simulations required only about 4 minutes of CPU time to run to completion, and proved to be robust, as no calculations aborted once a converged time step was used. The time step chosen was 0.001 s, which was sufficiently small such that temporal convergence was reached.

### Comparison Between Experimental Data and Code Simulations

As shown in Table 1, the simulation output compared favorably with data for the nine simulations—the code output was within 1 to 14% error, with eight of the nine showing less than 10% error. Because the experimental data has not yet been released, and because there is an ongoing analysis of the data, this paper includes the normalized $SO_2$ production rates and a comparison between experimental and calculated $SO_2$ production efficiencies. Nevertheless, these comparisons provide a useful canon by which to gauge MELCOR-H2 performance with the experimental data. Readers interested in the explicit data are referred to the relevant papers [16, 17].

The simulations showed that for a given molar flow rate, the higher the solution mole %, the higher the conversion efficiency. This trend was also reflected in the data, as shown in Table 1. Additionally, the simulations indicated that the acid conversion efficiency dropped as the molar flow rate increased. This is reasonable because faster molar flow rates should result in shorter fluid $r_t$ within the catalyst, thus decreasing the amount of reaction. An observation of Table 1 shows that the data also captured this trend.

Finally, Fig. 3 shows the normalized $SO_2$ production rate, which provides a direct indication of the expected production of $H_2$. The agreement between the simulations and data is very good.

**Table 1.** Comparison Between Experimental Data and Simulation Output.

| Solution Mole % | Normalized Acid Flow Rate | Conversion Efficiency: Difference Between Experimental Data and MELCOR-H2 Output (%) |
|---|---|---|
| 19 | 0.28 | 3.6 |
| 19 | 0.53 | -2.0 |
| 19 | 1.00 | -1.6 |
| 38 | 0.26 | 6.5 |
| 38 | 0.46 | 6.6 |
| 38 | 0.49 | 1.6 |
| 42 | 0.27 | 8.7 |
| 42 | 0.44 | 0.92 |
| 42 | 0.99 | -14.3 |

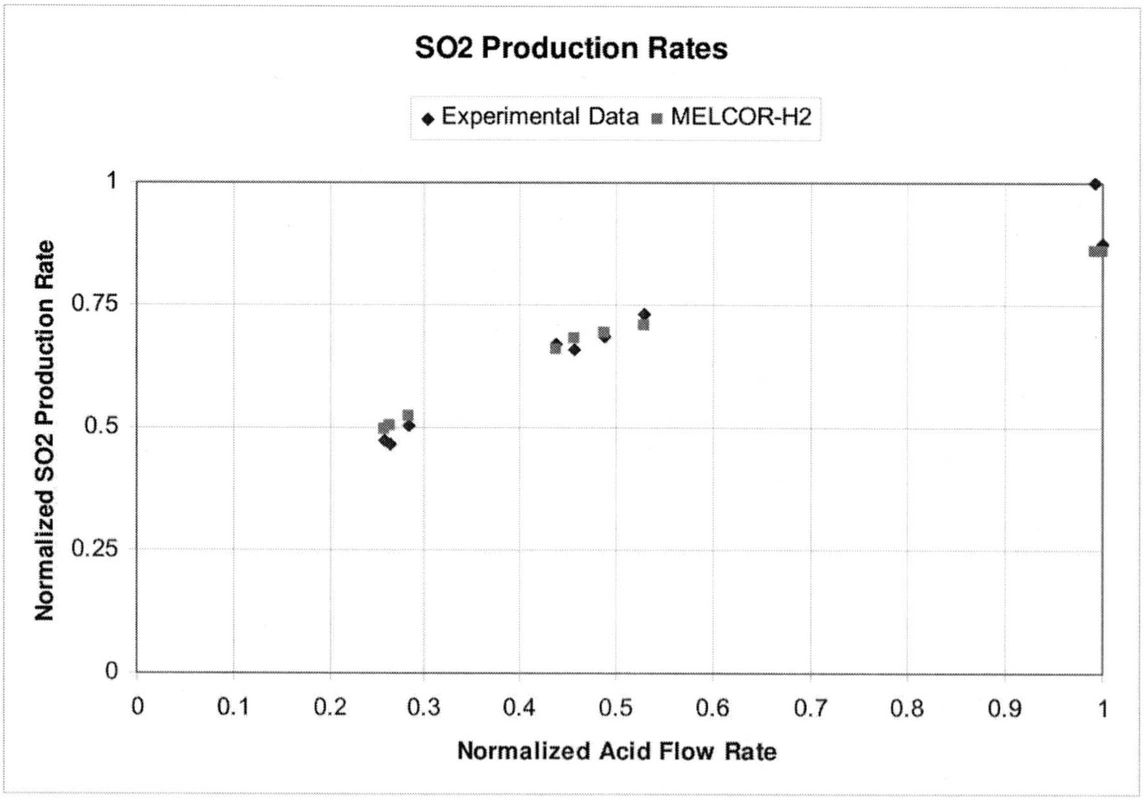


**Fig. 3. Normalized Experimental and Computed $SO_2$ Production Rates.**

## Conclusion

Sulfuric acid decomposition experiments conducted at Sandia National Laboratories were simulated using MELCOR-H2. The code output agreed quite well with experimental data. In addition, observed trends in the simulations were reflected by the experimental data; this included the relationship between solution mole % and conversion efficiency, as well as the relationship between molar flow rate and efficiency.

## *References*


1. L. C. Brown, G. E. Besenbruch, R. D. Lentsch, *et al*, "High Efficiency Generation of Hydrogen Fuels Using Nuclear Power", General Atomics, GA-A24285, Rev. 1 (December 2003).
2. Nariaki Sakaba, Seiji Kasahara, Hirofumi Ohashi, *et al*, "Hydrogen Production by Thermochemical Water-Splitting IS Process Utilizing Heat from High-Temperature Reactor HTTR", World Hydrogen Energy Conference, Lyon, France (June 2006).
3. Mohamed S. El-Genk and Jean-Michel Tournier, "Thermo-Chemical Generation of Hydrogen Using High Temperature Gas Power Cycle Modeling Recommendation in MELCOR", Institute for Space and Nuclear Power Studies and the Chemical and Nuclear Engineering Department, University of New Mexico (February 2006).
4. C. H. Oh, C. Davis, S. Shermal, *et al*, "International Nuclear Engineering Research Initiative Project at INL, ANL, and KAERI", ANS Thermal Hydraulics Division Newsletter, Research Briefs (Spring 2006).
5. R. O. Gauntt, J. E. Cash, R. K. Cole, *et al*, "MELCOR Computer Code Manuals. Vol. 1: Primer and Users' Guide Version 1.8.6", Sandia National Laboratories (September 2005).
6. Kevin J. Hogan, Karen Vierow, Shripad T. Revankar, *et al*, "Assessment of PBMR Analysis Using the MELCOR Code", ANS 2006 Annual Meeting, Reno, Nevada (June 2006).
7. Philip E. MacDonald, "Advanced Reactor, Fuel Cycle, and Energy Products Workshop for Universities", http://neri.inel.gov/universities_workshop/proceedings/pdfs/ngnp.pdf, Idaho National Engineering and Environmental Laboratory (March 4-5, 2004).
8. Sal B. Rodriguez, Randall O. Gauntt, Randy Cole, *et al*, "MELCOR Extensions for Simulation of Modular Power Cycles and Thermochemical Cycles for the Generation of Hydrogen via Nuclear Reactors", World Hydrogen Energy Conference, Lyon, France (June 2006).
9. Sal B. Rodriguez, Randall O. Gauntt, Randy Cole, *et al*, "Development of Design and Simulation Model and Safety Study of Large-Scale Hydrogen Production Using Nuclear Power", AIChE, Columbus, Ohio (November 2005).
10. Sal B. Rodriguez, Randall O. Gauntt, Randy Cole, *et al*, "MELCOR Modification for Large-Scale Hydrogen Production Using Nuclear Thermochemical Cycles", ANS (November 2005).
11. Shripad T. Revankar, Seungmin Oh, Sal Rodriguez, "Simplified Model to Couple SI Cycle to Nuclear Heat Transport System", ANS 2006 Annual Meeting, Reno, Nevada (June 2006).

12. Shripad T. Revankar and Seungmin Oh, "Simplified Sulfur Iodine Chemical Water Splitting Process and Heat Exchanger Equations for Implementation in MELCOR Interface", Purdue University (December 19, 2005).
13. David Louie, Sal B. Rodriguez, Randy Cole, *et al*, "Report on Implementing the SI Chemistry Model into MELCOR for the Sandia Hydrogen Generation Project", Omicron Safety and Risk Technologies (April 17, 2006).
14. Rodriguez, Sal, Randall O. Gauntt, Randy Cole, *et al*, "MELCOR-H2: A Modular, Generalized Tool for the Dynamic Simulation and Design of Fully-Coupled Nuclear Reactor/Hydrogen Production Plants", 2006 ANS Winter Meeting, Albuquerque, NM.
15. Rodriguez, Sal, Randall O. Gauntt, Randy Cole, *et al*, "Modeling of a Z-IFE Hydrogen Plant Using MELCOR-H2", November 2006 TOFE, Albuquerque, NM.
16. Gelbard, Fred, Robert Moore, Edward Parma, *et al*, "Sulfuric Acid Decomposition Experiments for Thermochemical Hydrogen Production from Nuclear Power", 2007 American Nuclear Society Annual Meeting, Boston.
17. Gelbard, Fred, Robert Moore, Edward Parma, *et al*, "Sulfuric Acid Decomposition Experiments", Sandia National Laboratories, February 2007.
18. Huang, C. and A. T-Raissi, "Analysis of Sulfur-Iodine Thermochemical Cycle for Solar Hydrogen Production. Part I: Decomposition of Sulfuric Acid," Solar Energy 2005, 78:632-646.
19. S. Ergun, "Fluid Flow Through Packed Columns", Chem. Eng. Prog., 1952, 48:89-94.

## Acknowledgements

The authors are deeply grateful to the Sandia National Laboratories Sulfuric Acid Decomposition Experiment Team, including Fred Gelbard, Robert C. Moore, Edward J. Parma, Milton E. Vernon, and Howard B. J. Stone.

## High Temperature Gas-Cooled Reactor Coupled with High Temperature Steam Electrolysis

Chang H. Oh[1*], Cliff Davis[1], Robert Barner[2] Eung S. Kim[1], Steven Sherman[1], and R. Vilim[3]

1. *Idaho National Laboratory, P.O. Box 1625, Idaho Falls, ID 83415-3890, chh@inel.gov*
2. *Summer Intern, Texas A&M University, Department of Nuclear Engineering 3133 TAMU, College Station, TX 77843-3133*
3. *Argonne National Laboratory, 9700 South Cass Ave., Argonne, IL 60439*

## ABSTRACT

The US Department of Energy is investigating the use of high-temperature gas-cooled reactors (HTGR) to produce electricity and hydrogen. Although the hydrogen production processes using the nuclear energy are in an early stage of development, coupling hydrogen plant to HTGR requires both efficient heat transfer and adequate separation of the facilities to assure that off-normal events in the production facility do not impact the nuclear plant.

In anticipation of the design, development and procurement of an advanced power conversion system for HTGR, this study was initiated to identify the major design and technology options and their tradeoffs in the evaluation of power conversion unit (PCU) coupled to the hydrogen plant. In this study, we investigated a number of design configurations and performed thermal hydraulic analyses using various working fluids and various conditions. Also using the high temperature steam electrolysis process (HTSE), we calculated the energy requirement for the operation of the HTSE. The balance of the plant of the HTSE was developed and the HTGR was modeled to be coupled with the HTSE. This paper presents a portion of results obtained from this study.

## I. INTRODUCTION

The Department of Energy and the Idaho National Laboratory are developing a Next Generation Nuclear Plant (NGNP) to serve as a demonstration of state-of-the-art nuclear technology. The purpose of the demonstration is two fold: 1) efficient low cost energy generation and 2) hydrogen production. Although a next generation plant could be developed as a single-purpose facility devoted to hydrogen production, early designs are expected to be dual-purpose. While hydrogen production and advanced energy cycles are still in its early stages of development, research towards coupling a high temperature reactor, electrical generation and hydrogen production is under way. Many aspects of the NGNP must be researched and developed in order to make recommendations on the final design of the plant. Parameters such as working conditions, materials, stresses, cycle components, working fluids, coupling of the hydrogen production plant and power conversion unit (PCU) configurations must be understood.

In this study, we investigated various PCU configurations coupled to a High Temperature Steam Electrolysis (HTSE) plant by means of an intermediate heat transport loop (IHTL). The key issues that are addressed in this study are PCU configuration options, coupling of the HTSE to the reactor, various working fluids in the PCU and IHTL, and efficiency. However, in this paper we present (1) coupling of the HTSE to the reactor and (2) the overall efficiency of the coupled VHTR/HTSE plant.

The commercial process code HYSYS [1] was used to model design configurations of the PCU coupled to a HTSE plant. An IHTL was used for transporting heat to the HTSE plant. This IHTL was taken from configuration 6 from a report by [2]. The HTSE plant was adapted from work by [3]. This same IHTL and HTSE were used for all configurations and parametric studies in this study.

Helium, $CO_2$, and molten salt were studied to determine the best working fluid in terms cycle efficiency and development cost. Helium is a well understood fluid and has been used in numerous studies pertaining to nuclear power. $CO_2$ has been slow in developing due to material concerns with the fluid. $CO_2$ does possess some advantages over helium such as a higher density allowing for smaller velocities than helium for the same pressure drops [4]. Despite the lower specific heat, the volumetric flow rates are smaller for $CO_2$ than for a helium cycle generating equivalent power. Therefore, the turbomachinery sizes are smaller for $CO_2$. Copsey et al. [5] used the nitrogen-helium mixture for the working fluid in a combined cycle.

## 2. METHODS

This section describes the methods used to perform the thermal-hydraulic and efficiency evaluations. The method used to estimate the overall cycle efficiency, process optimization, and coupling of VHTR and HTSE are described below.

### 2-1. Efficiency

The efficiency of each proposed configuration was estimated using HYSYS [1], a process optimization code used in the chemical and oil industries.

The PCU cycle efficiency, $\eta_{PCU}$, used in this study is defined as:

$$\eta_{PCU} = \frac{\text{Electric power output}}{\text{Reactor thermal power - H}_2\text{ process power}} = \frac{\Sigma W_T - \Sigma W_C - W_S - \Sigma W_{CIR}}{Q_{th} - Q_{H2}}, \quad (1)$$

where $\Sigma W_T$ is the total turbine workload, $\Sigma W_C$ is the total compressor workload, $W_S$ is the plant stationary load, $\Sigma W_{CIR}$ is the circulator workload in the primary, intermediate, and, if present, tertiary loops, $Q_{th}$ is the reactor thermal power, and $Q_{H2}$ is the power supplied through the PHX to the hydrogen generating plant. For the efficiency calculations, we report the overall cycle efficiency, which is defined as

$$\eta_{overall} = \frac{\Sigma W_T - \Sigma W_C - \Sigma W_{CIR} - \Sigma W_{H2} - Q_{EL} + Q_{H2}}{Q_{th}}, \quad (2)$$

where $\Sigma W_T$ is the total turbine workload, $\Sigma W_C$ is the total compressor workload, $W_{CIR}$ is the circulator workload in the primary and secondary side, $\Sigma W_{H2}$ is the workload in the HTSE plant, $Q_{EL}$ is the power supplied to the electrolyzer, $Q_{H2}$ is the lower heating value of the produced hydrogen and $Q_{th}$ is the reactor thermal power.

The polytropic efficiency, rather than the isentropic efficiency, is used for representing the efficiency of the turbomachinery. The equations for the expansion and compression processes in a perfect gas are taken from Saravanamuttoo et al. [6]. For an expansion, the efficiency is calculated from

$$\frac{T_{0,ex}}{T_{0,in}} = \left(\frac{P_{0,ex}}{P_{0,in}}\right)^{\left(\frac{R}{C_P}\eta_{p,e}\right)}, \tag{3}$$

where R is the gas constant, $C_p$ is the specific heat, $\eta_{p,e}$ is the turbine polytropic efficiency, $T_0$ is the stagnation temperature, and $P_0$ is the stagnation pressure. Subscripts *ex* and *in* refer to exit gas and inlet gas, respectively. For a compression, the efficiency is calculated from

$$\frac{T_{0,ex}}{T_{0,in}} = \left(\frac{P_{0,ex}}{P_{0,in}}\right)^{\left(\frac{R}{C_P}\frac{1}{\eta_{p,c}}\right)}. \tag{4}$$

HYSYS was used to develop an input model for each configuration and working fluid and to optimize the cycle efficiency. HYSYS uses the Peng-Robinson [7] equation of state to determine the properties of the working fluids. However, HYSYS does not have thermal properties for molten salts. Therefore, the physical and thermal properties of a molten salt were input in a tabular form.

The efficiencies of the turbine and compressors were assumed to be 92% and 90%, respectively. The pressure ratio, which is defined as the outlet pressure from the high pressure compressor (HPC, 1$^{st}$ compressor after precooler in Figure 3) divided by the inlet pressure to the low pressure compressor (LPC), was varied to optimize the overall cycle efficiency. Cooling was applied between compressors to reduce the power consumed by the HPC. Cooler components were used to simulate the heat loss and differential pressure along the hot and cold legs of the IHTL.

### 2-2. Process Optimization

For this calculation, the process optimization was carried out using HYSYS process modeling software. HYSYS has an optimization tool that is built in to integrate the simulation model of the system. The optimization program searches for the maximum value of a given objective function subject to a number of imposed constraints. Figure 1 shows 2-dimensional design spaces with defined regions. The goal is to maximize f (x1, x2, x3) where x1, x2 are independent variables such as mass flow, pressure and etc. For the optimization, the x variables are manipulated within a specified range of a lower and upper bound. The regions in Figure 2 are defined by a feasible design space within functional constraints and an infeasible design space outside of the constraint boundaries [8]. Functional constraints are material and energy balances, for example, positive pressure drop in every stream in the direction of flow, positive power in turbine and compressor, the temperature requirement at the inlet and outlet of the heat exchangers, etc. The function, f, is the objective function which is defined as the overall plant busbar efficiency above. Constant values of the objective function define contours on the design space. Figure 2 shows design space of 2-D design vector showing optimal design points. First, the initial calculation should be in the feasible design space and the search continues towards a direction in the design space until a maximum is reached. A new direction is found and the search continues in that direction as long as the objective function value increases. Once it reaches the contour, the search continues towards the optimal design point that is the optimum solution. However, if the optimal design point is outside of the feasible design space, the closest contour to the optimal solution that coincides with the constraint boundary can be the optimal solution within the constraints specified.

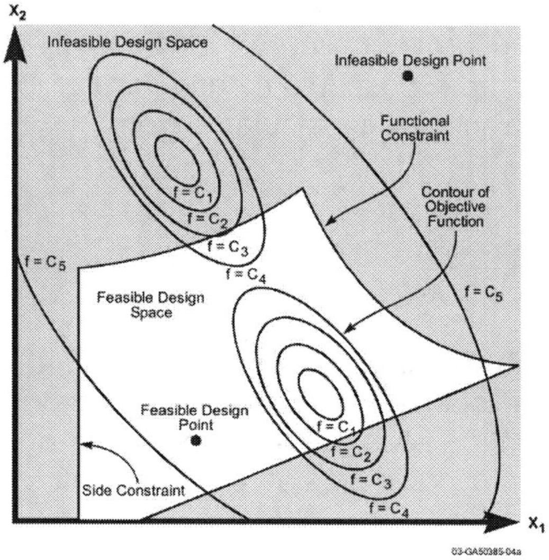


Figure 1. Design space with designed regions.

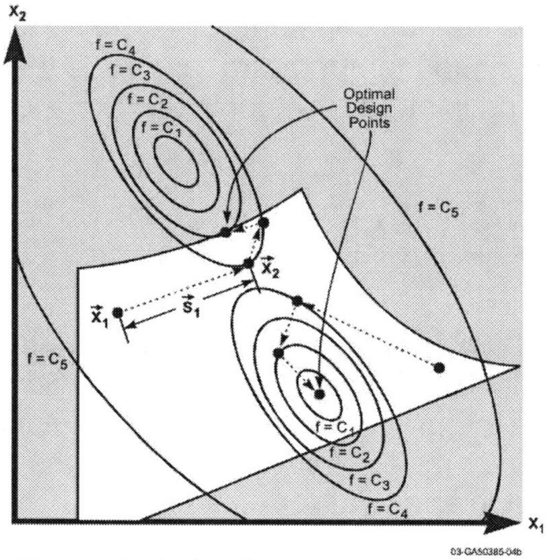


Figure 2. Potential optimal solutions.

### 2-3 Coupling of VHTR and HTSE

The following description is how VHTR (600 $MW_t$) is coupled with HTSE. The baseline plant configuration was based on Figure 3. In this configuration an indirect cycle was used with the reactor coupled to the secondary side by means of an intermediate heat exchanger (IHX) as recommended by the Independent Technology Review Group [9]. The reactor was assumed to produce 600 MW of thermal power with a 900 °C outlet temperature with a nominal pressure of 7MPa and use helium coolant on the primary side. The nominal rise in fluid temperature across the core was assumed to be 400 °C, based on the point design [10]. The reactor pressure drop and the primary side pressure drop of the IHX are assumed to be 50kPa.

The power conversion unit (PCU) consists of a recuperated Brayton cycle. Referring to Figure 3, the secondary fluid exits the IHX and the flow is split with most of the flow entering the PCU. The rest of the flow, approximately 10%, goes to the HTLHX. Upon entering the PCU the fluid enters the turbine for expansion. After leaving the turbine the fluid passes through the hot side of the recuperator and is further cooled by a precooler before entering the low pressure compressor. The fluid is the cooled aging by an intercooler before entering the high pressure compressor. The fluid leaves the high pressure compressor and enters the cold side of the recuperator before combining with the fluid from the HTLHX and returning to the IHX.

The reactor was coupled to the hydrogen processing facility by means of an intermediate heat transport loop (IHTL). The IHTL was placed in parallel with the PCU as seen in Figure 3. It was assumed that 50MW of thermal power are transported to the IHTL. A circulator is needed in the secondary side account for pressure losses in the heat transport loop heat exchanger (HTLHX). Estimations of the required separation distance between the nuclear and hydrogen process plant vary considerable. For example, Sochet et al. [11] recommended 500 m for the High-Temperature Reactor while Smith et al. [12] recommended a separation distance of from 60 to 120 m for the NGNP and the hydrogen production plant. For this analysis, a nominal value of 90 m was used.

The coupling of the IHTL to the hydrogen processing facility was accomplished by means of three process heat exchangers (PHX). Figure 4 details the configuration of the 3 PHX's in the IHTL. Two heat exchangers in parallel are followed by one heat exchanger in series. This configuration was chosen to deliver high inlet temperatures on the hot side of the first two heat exchangers where high cold side outlet temperatures are needed for the hydrogen production facility. The third heat exchanger however, does not require high cold side outlet temperatures and the hot side inlet temperature from the outlet of the prior heat exchangers is sufficient for heating the cold side fluid.

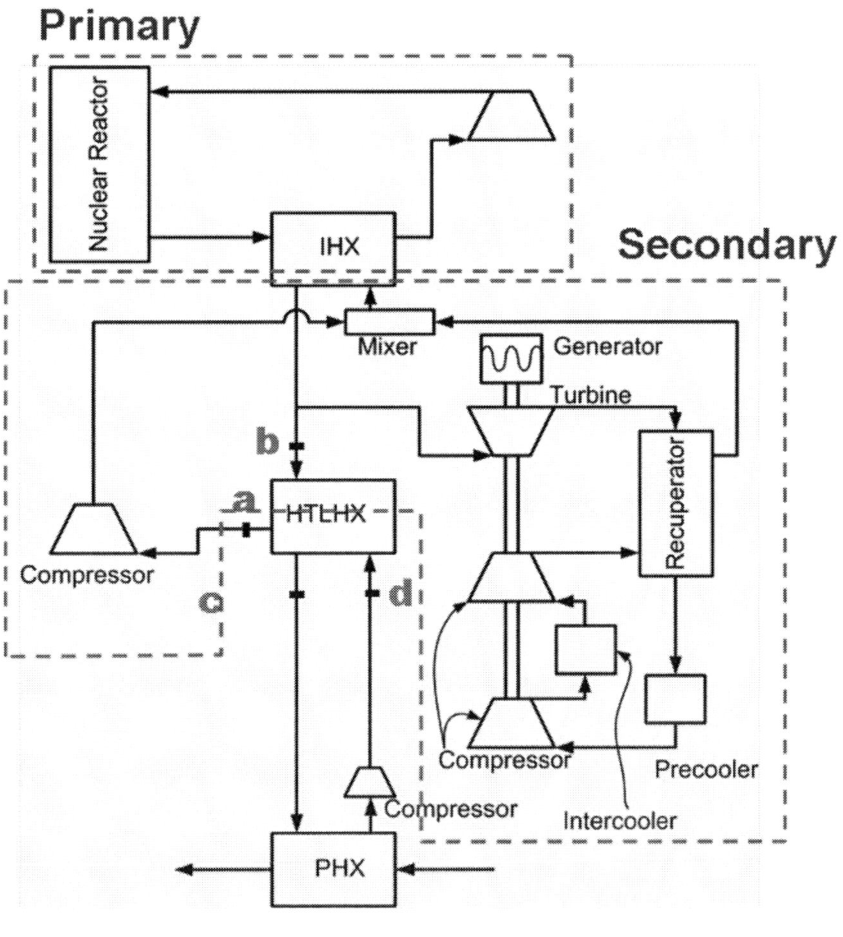


Figure 3. Baseline plant configuration for electrical and hydrogen production

## Ternary

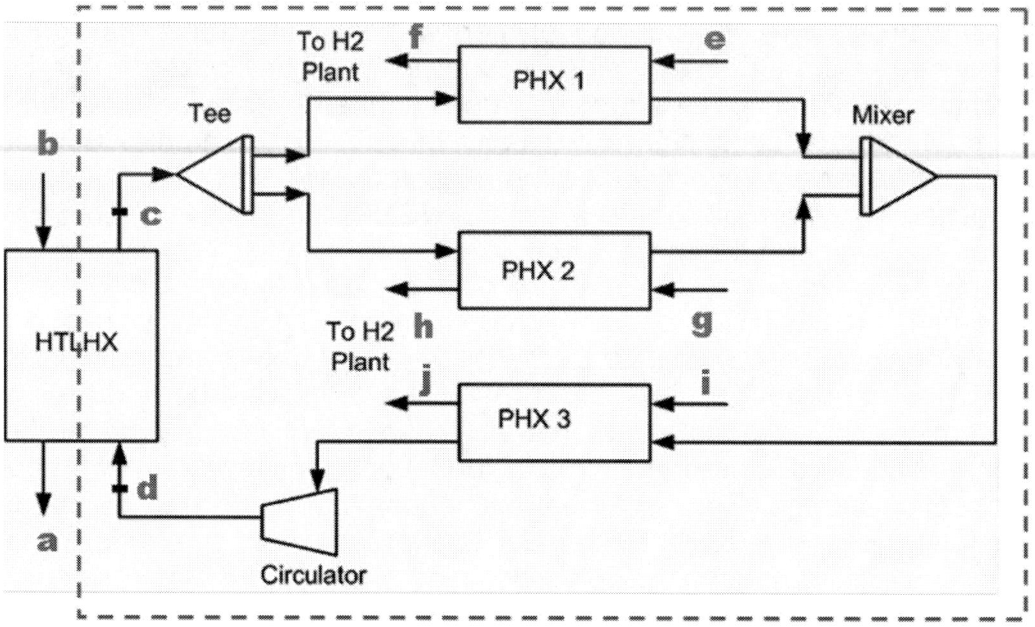


Figure 4. IHTL process heat exchanger configuration

The details of the HTSE are shown in Figure 5. The process water enters on the left. The water is then pumped up to the operating pressure of 5 MPa. The efficiency of the pumps and circulators is assumed to be 75%. This water is then combined with water condensate returned from the hydrogen/water product stream. This stream then enters the low temperature recuperator. The pressure drop through the heat exchangers is assumed to be 20 kPa. From there the steam is further heated by PHX 3. Upon leaving PHX 3 the steam is mixed with hydrogen from the product stream by a recirculator which works to overcome the pressure drops in the system. A mole fraction of 90% water and 10% hydrogen is maintained in this model. This hydrogen helps to maintain reducing conditions at the electrolysis stack to prevent oxidation. The mixed stream then enters the high temperature recuperator which takes advantage of the high temperature outlet from the electrolysis stack. The hydrogen/water stream is then heated to the operating temperature for the electrolysis stack, in this case 827 °C, in PHX 1.

The electrolyzer has another inlet stream that contains the sweep gas. This is used to sweep away the oxygen from the electrolysis process. A steam sweep gas is used in this model and enters the plant in the middle-bottom of Figure 5. It is first pumped up to operating pressure and then heated in a heat exchanger using the hot sweep outlet from the electrolyzer. Before it enters the electrolysis stack it is heated to the operating temperature of 827 °C in PHX 2.

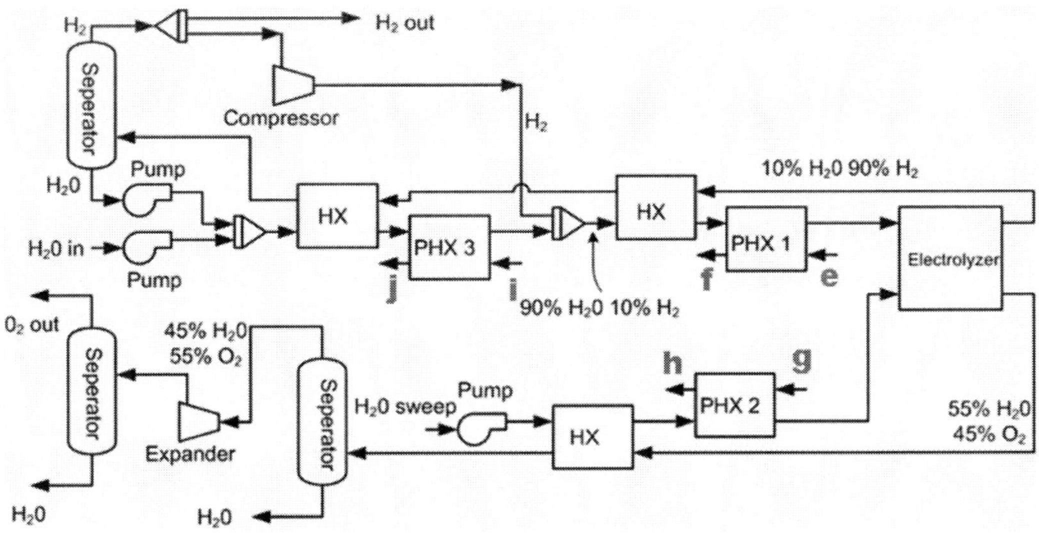


Figure 5. Schematic of HTSE process.

Upon leaving the electrolyzer, the product stream is 90% hydrogen and 10% water. This then passes through the high and low temperature recuperators. The steam condensate is then separated from the hydrogen and recycled back into the system. After leaving the electrolyzer, the sweep gas enters the recuperator to preheat the sweep inlet. The sweep outlet contains approximately 55% water and 45% oxygen. It is then partially separated before entering an expander to recuperate some of the power. The expander has an efficiency of 80%. Finally the sweep stream is further separated and high purity oxygen and water are produced.

### 2-4 Electrolyzer Model

In the electrolyzer model [13] the oxygen stream produced at the anode is assumed to mix with a sweep gas stream that is introduced at the anode. The combined stream then exits the electrolyzer. The hydrogen stream produced at the cathode is assumed to mix with a feed stream that is introduced at the cathode. The feed stream is composed of water vapor to be electrolyzed, hydrogen gas for maintaining reducing environment, and possibly an inert gas, presently assumed to be nitrogen.

A. Energy Equation

An energy balance on the electrolyzer gives

$$\sum_i \dot{n}_{P-i}\, H_{P-i}(T_P, P) = \sum_i \dot{n}_{R-i}\, H_{R-i}(T_R, P) + Q + W \tag{5}$$

where

| | | |
|---|---|---|
| $\dot{n}$ | = | species mole flow rate, |
| H | = | enthalpy per mole, |
| Q | = | rate of heat transfer to the electrolyzer, |

| | | |
|---|---|---|
| $W$ | = | rate of electrical work supplied to the electrolyzer, |
| $T$ | = | temperature, |
| P | = | pressure, |

and where we have used subscripts $R$ for reactants and $P$ for products. Their mass flowrates are defined

$$
\begin{aligned}
& m_{H_2O-o-cath}\, h_{H_2O}\left(T_o,P\right)+m_{H_2-o-cath}\, h_{H_2}\left(T_o,P\right)+m_{N_2-o-cath}\, h_{N_2}\left(T_o,P\right) \\
& \quad m_{O_2-o-anode}\, h_{O_2}\left(T_o,P\right)+m_{sweep-o-anode}\, h_{sweep}\left(T_o,P\right)= \\
& \qquad m_{H_2O-i-cath}\, h_{H_2O}\left(T_i,P\right)+m_{H_2-i-cath}\, h_{H_2}\left(T_i,P\right)+m_{N_2-i-cath}\, h_{N_2}\left(T_i,P\right) \\
& \qquad\quad m_{O_2-i-anode}\, h_{O_2}\left(T_i,P\right)+m_{sweep-i-anode}\, h_{sweep}\left(T_i,P\right)+Q+W
\end{aligned} \tag{6}
$$

where

| | | |
|---|---|---|
| $m$ | = | species mass flow rate (kg/s), |
| $h$ | = | specific enthalpy (joules/kg), |

and where subscripts $i$ and $o$ represent inlet and outlet, respectively.

B. Species Mole and Mass Flowrates

The species mole flowrates entering and leaving the electrolyzer are related to the current density through the relationships

$$
\begin{aligned}
& \dot{n}_{H_2O-o-cath}=\dot{n}_{H_2O-i-cath}-\frac{iA}{2F} \\
& \dot{n}_{H_2-o-cath}=\dot{n}_{H_2-i-cath}+\frac{iA}{2F} \\
& \dot{n}_{O_2-o-anode}=\dot{n}_{O_2-i-anode}+\frac{iA}{4F} \\
& \dot{n}_{sweep-o-anode}=\dot{n}_{sweep-i-anode} \\
& \dot{n}_{N_2-o-cath}=\dot{n}_{N_2-i-cath}
\end{aligned} \tag{7}
$$

where

| | | |
|---|---|---|
| $i$ | = | current density (amps/$m^2$), |
| $A$ | = | electrode surface area, ($m^2$) and |
| $F$ | = | Faradays constant. |

The species mass flowrates and mole flowrates are related as follows: For an individual species

$$\dot{m}_{k-o}=A_k\ \dot{n}_{k-o} \text{ and } \dot{m}_{k-i}=A_k\ \dot{n}_{k-i}\ ,\ k = H_2O,\ H_2\ ,\ O_2, \text{ and } N_2. \tag{8}$$

where $A_k$ is the atomic weight of species $k$ in kg per mole and subscript $o$ is the outlet and $i$ is the inlet.

C. Cell Voltage and Electrical Work

The voltage drop across the electrolyzer is the sum of the electrode Nernst potential and the resistance of the cell. In estimating the resistance, the activation and the concentration overpotentials are lumped in with the cell internal resistance. The cell voltage is then assumed given by

$$V_{cell} = V_N + i \cdot r \tag{9}$$

Where

| | | |
|---|---|---|
| $V_N$ | = | is the Nernst potential, and |
| $r$ | = | is the area-specific cell resistance (ohms-m$^2$). |

The electrical work done in the cell is

$$W = V_{cell} \cdot i \cdot A \; . \tag{10}$$

The active species giving rise to the Nernst potential satisfy the chemical balance equation

$$H_2O\,(g) \rightarrow H_2\,(g) + \frac{1}{2} O_2\,(g) \, . \tag{11}$$

The change in Gibbs free energy for this reaction carried out at temperature T and pressure P is

$$\ddot{A}G(T,P) = \ddot{A}G_f(T,P) + RT \ln \left[ \frac{f_{H_2}\, f_{O_2}^{\frac{1}{2}}}{f_{H_2O}} \right] \tag{12}$$

where $f$ is the molar fraction of a species and $\ddot{A}G_f(T,P)$ is the Gibbs free energy in forming the products at temperature $T$ and pressure $P$ minus the same for the reactants, that is,

$$\ddot{A}G_f(T,P) = G_{f-H_2}(T,P) + 1/2\; G_{f-O_2}(T,P) - G_{f-H_2O}(T,P) \, . \tag{13}$$

where $G_{f\text{-}i}\,(T,P)$ ) is the Gibbs free energy on a per mole basis of forming species $i$ at conditions $T$ and $P$. In turn $\ddot{A}G_f(T,P)$ is written in terms of $\ddot{A}G^0{}_f(T) = \ddot{A}G_f(T,P_{STD})$ where $P_{STD}$ = 0.101 MPa. Then setting the change in Gibbs free energy equal to the electrical work done the voltage developed by the cell is

$$V_{Nernst} = \frac{-1}{2F}\left[\ddot{A}G_f^{\,0}(T) + RT\ln\left[\left(\frac{f_{H_2}\, f_{O_2}^{\frac{1}{2}}}{f_{H_2O}}\right)\left(\frac{P}{P_{STD}}\right)^{\frac{1}{2}}\right]\right] \tag{14}$$

where $P_{STD}$ = 0.101 MPa and $P$ is the cell pressure.

The mole fractions at any point in the electrolyzer are related to the molar mass flowrates at that point through

$$f_{H_2O-cath} = \frac{\dot{n}_{H_2O}}{\dot{n}_{H_2O} + \dot{n}_{H_2} + \dot{n}_{N_2}} \qquad f_{H_2-cath} = \frac{\dot{n}_{H_2}}{\dot{n}_{H_2O} + \dot{n}_{H_2} + \dot{n}_{N_2}}$$
$$f_{O_2-anode} = \frac{\dot{n}_{O_2}}{\dot{n}_{sweep} + \dot{n}_{O_2}} \qquad f_{sweep-anode} = \frac{\dot{n}_{sweep}}{\dot{n}_{sweep} + \dot{n}_{O_2}}. \tag{15}$$

The current density and active cell area are then specified, yielding the total operating current. Care must be taken to insure that the specified inlet gas flow rates and total cell current are compatible. The minimum required inlet steam molar flow rate is the same as the steam consumption rate, given by:

$$\dot{N}_{i,H2O,\min} = \Delta\dot{N}_{H2O} = \frac{I}{2F} N_{cells} = \frac{i\,A_{cell}}{2F} N_{cells} = \Delta\dot{N}_{H2} \tag{16}$$

which is of course also equal to the hydrogen production rate.

Once the total and per-cell hydrogen production rates are known, the outlet flow rates of hydrogen and steam on the cathode side and oxygen on the anode side can be determined. The flow rates of any inert gases, the anode-side sweep gas, and any excess steam or hydrogen are the same at the inlet and the outlet. Once all these flow rates are known, the summations in Eqn. (5) can be evaluated. The product summation must be evaluated initially at a guessed value of the product temperature, $T_P$.

## 3. RESULTS

There are three flow loops in the coupled HTGR/HTSE plant: the primary loop contains VHTR, IHX, and a compressor and the secondary loop involves components in the PCU while the ternary loop involves all the components in the HTSE. The preliminary results on the working fluids in the primary, secondary, and ternary are summarized below:

Table 1 shows the overall efficiency of the coupled VHTR/HTSE plant using Eqn. (2). These calculations are based on a 600MW$_t$ from the VHTR and a 50 MW thermal energy is transferred to the HTSE through HTLX (streams c and d) in Figure 4. The melting point of Flinak is 454$^0$C and therefore a higher temperature of 454$^0$C was used in the process side in the HTSE where Flinak, a molten salt, was used as one working fluid of two other fluids used in this scoping analysis.

Table 1. Cycle overall efficiency

| Primary | Secondary | Ternary | Efficiency (% |
|---|---|---|---|
| He | He | He | 44.12 |
| He | He | CO2 | 44.47 |
| He | He | Flinak | 46.13 |
| He | CO2 | He | 43.35 |
| He | CO2 | CO2 | 43.5 |
| He | CO2 | Flinak | 43.41 |
| Flinak | He | He | 45.25 |
| Flinak | He | CO2 | 45.5 |
| Flinak | He | Flinak | 47.24 |
| Flinak | CO2 | He | 44.83 |
| Flinak | CO2 | CO2 | 45.06 |
| Flinak | CO2 | Flinak | 45.39 |

Figure 6 shows results of electrical work done on the electrolyzer using Eqns (10) and (14). The area, A, defined in Eqn (10) is the surface area per one cell. Therefore, in order to calculate the electrical work, number of cells are multiplied in Eqn (10). A shown in Figure 6, the increase of the current density reduces number of cells and increases the electrical work on the electrolyzer.

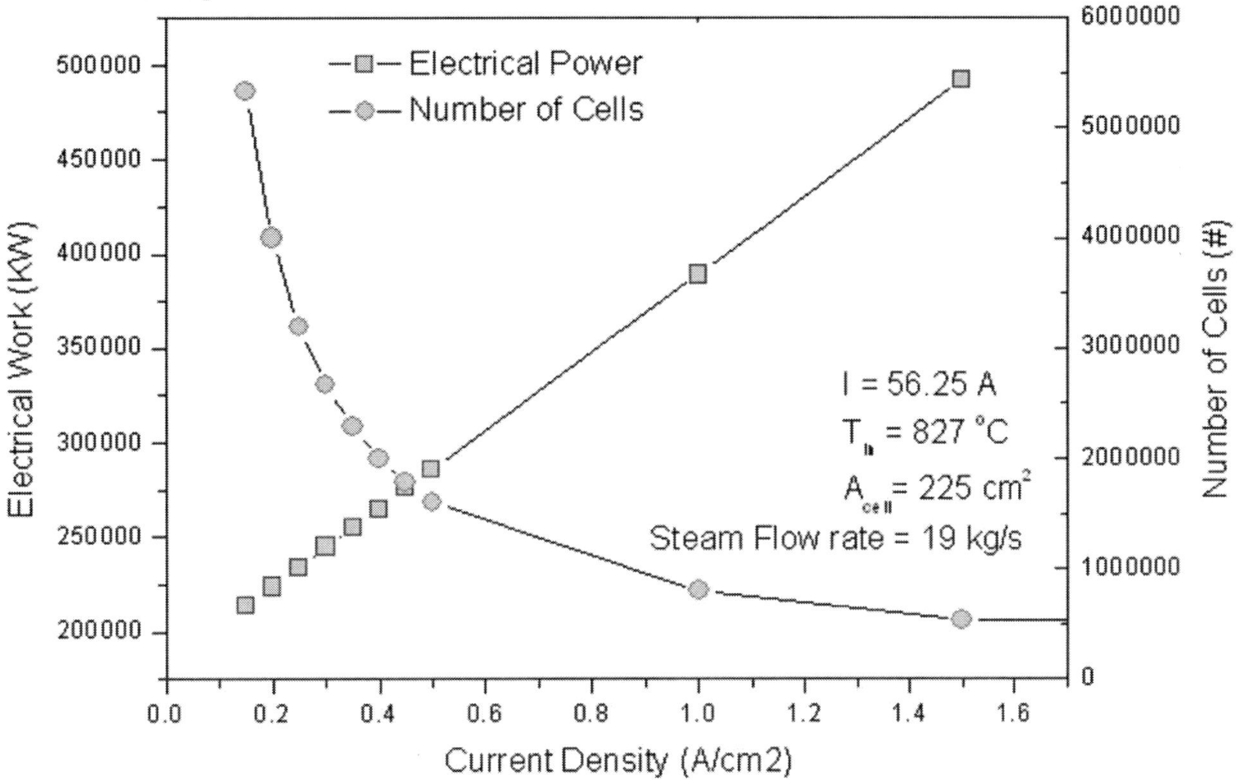


Figure 6. Required electrical work done on the electrolyzer as a function of current density and number of cells.

## 4. CONCLUSIONS

A reference design of a coupled nuclear reactor and hydrogen production plant has been established. Analysis of the full power operating conditions was performed with the HySYS code combined with a specialized model developed for the electrolytic cell. Important phenomena relating to accurate prediction of coupled plant efficiency were modeled. Included was the temperature drop from hot to cold side in heat exchangers, the ohmic and overpotential losses in the electrolytic cell, irreversibilities in compressors, turbines, and pumps, and the pressure drop in heat exchangers and long pipe runs.

Of various combinations of working fluids investigated, the greatest coupled plant efficiency is obtained with a molten salt in the primary and ternary flow loops and helium in the secondary loop. This plant shows an overall efficiency of 47%, which is approximately 3% higher than the same plant with helium in all thee flow loops. The increased efficiency arises from the reduced pumping losses associated with a liquid coolant compared to a gas. Choosing between a gas and a molten salt in the primary system has implications for the core neutronics not addressed in this paper.

In future work, we will perform a cost analysis based on the component sizes estimated in the present work. Additionally, we have recently identified a plant configuration that has comparable efficiency but may have reduced cost with respect to the heat transport loop that supplies power to the hydrogen plant.[14] This may merit a comparative economic analysis.

Acknowledgement: This work was supported through the U.S. Department of Energy's Power Conversion Program and Nuclear Hydrogen Initiative Program under DOE Idaho Operations Office Contract DE-AC07-99ID13727.

## REFERENCES

[1] Aspen Technology, HYSYS Process Version 2.2.2, www.aspentech.com,

[2] Davis, C. B., C. H. Oh, R. B. Barner, S. R. Sherman, and D. F. Wilson, Thermal-Hydraulic Analyses of Heat Transfer Fluid Requirements and Characteristics for Coupling a Hydrogen Production to a High-Temperature Nuclear Reactor, INL/EXT-05-00453, June 2005.

[3] Stoots, C. M., "Engineering Process Model for High-Temperature Electrolysis System Performance Evaluation," AICHE conference, Cincinnati OH, October 2005.

[4] Perry, R. H. and D. W. Green, Perry's Chemical Engineers' Handbook, Seventh Edition, McGraw-Hill, 1997.

[5] Copsey, B., M. Lecomte, G. Brinkmann, A. Capitaine, and N. Deberne, The Framatome ANP Indirect-Cycle Very High Temperature Reactor, Proceedings of ICAPP '04, Paper 4201, June 2004.

[6] Saravanamuttoo, H. et al., 1996, Gas Turbine Theory, Fifth Edition, Prentice Hall.

[7] Peng, D.Y. and Robinson, D.B., 1976, "A Two Constant Equation of State," I.E.C. Fundamentals, 15, pp. 59-64.

[8] Mckellar, M.G., Optimization of a Household Refrigerator Considering Alternative Refrigerants, Ph.D. Thesis, Department of Mechanical Engineering, Purdue University, August 1992.

[9] Independent Technology Review Group, Design Features and Technology, Uncertainties for the Next Generation Nuclear Plant, INEEL/EXT-04-01816, 2004.

[10] MacDonald, P. E., J. W. Sterbentz, R. L. Sant, P. D. Bayless, R. R. Schultz, H. D. Gougar, R. L. Moore, A. M. Ougouag, and W. K. Terry, 2003, NGNP Preliminary Point Design – Results of the Initial Neutronics and Thermal-Hydraulic Assessments, INEEL/EXT-03-00870, July 2003.

[11] Sochet, I., J. L. Rouyer, and P. Hemmerich, 2004, "Safe Hydrogen Generation by Nuclear HTR," Paper 4261, Proceedings of ICAPP '04, Pittsburgh, PA, USA, June 13-17.

[12] Smith, C., S. Beck, and B. Galyean, 2005, An Engineering Analysis for Separation Requirements of a Hydrogen Production Plant and High-Temperature Nuclear Reactor, INL/EXT-05-00137 Rev 0, March 2005.

[13] Oh, C.H., C.B. Davis, J. Han, R. Barner, S. Sherman, R. Vilim, Y.J. Lee, and W.J. Lee, INEEL, 2006, HyPEP FY06 Report: Models and Method, INEEL/EXT-06-11725, September 2006.

[14] R.B. Vilim, "Power Requirements at the VHTR/HTE Interface for Hydrogen Production," Proceedings of ICAPP 2007 Nice, France, May 13-18, 2007.

# MOLTEN-PHASE HYDROLYSIS STAGE ANALYSIS AND EXPERIMENTS FOR THE CALCIUM BROMINE THERMOCHEMICAL CYCLE


Richard D. Doctor **
C. B. Panchal
Steven A. Lottes
Robert W. Lyczkowski and
Jianhong Yang,

** To whom correspondence should be sent



Key Words: hydrogen, nuclear, thermochemical, calcium bromide

Prepared for Presentation at the AIChE National Meeting, Houston, TX April 22-26, 2007




Unpublished

# Molten-Phase Hydrolysis Stage Analysis and Experiments for the Calcium Bromine Thermochemical Cycle

Richard D. Doctor, C.B. Panchal, Steven A. Lottes, Robert W. Lyczkowski, and Jianhong Yang
*Argonne National Laboratory, Argonne, IL 60439 USA*

## Abstract

The goal of the United States Department of Energy Nuclear Hydrogen Initiative as linked with the Generation IV Nuclear Energy Systems Initiative for Gas Reactor Deployment is to develop a cost-effective, proliferation-resistant, low-greenhouse-gas emissions, and sustainable, nuclear-based energy supply system [1]. The calcium-bromine cycle under development at Argonne National Laboratory combines both experimental and modeling studies of a novel continuous "hybrid" cycle for hydrogen production, where "hybrid" means that both nuclear heat and electricity are employed. Engineering the calcium-bromine cycle for continuous operation should facilitate its practical development since there will be an inherent advantage to using components and materials which will operate in a constant, non-cycling chemical and thermal environment. This paper focuses on the first and important calcium bromide hydrolysis stage to generate hydrogen bromide, which when split by electrolysis, produces hydrogen.

## Introduction

The calcium-bromine (Ca-Br) cycle is particularly attractive because nearly one-half the required thermodynamic energy for water splitting is delivered as nuclear reactor heat at around 1023 K (750 °C). It is envisioned that this temperature is low enough to facilitate the engineering of materials, while being high enough to have few steps in the cycle. The work at Argonne National Laboratory focuses on two special aspects of the Ca-Br cycle:

- Determining the feasibility of hydrogen ($H_2$) generation from hydrogen bromide (HBr).
- Determining the feasibility of calcium bromide ($CaBr_2$) hydrolysis in a continuous mode recognizing that there is a eutectic phase with calcium oxide (CaO) [2].

The three reactions in the hybrid Ca-Br cycle begin with the hydrolysis reaction that now appears to be a second order hydrolysis reaction in $CaBr_2$ involving the formation of a $CaBr_2$ and CaO complex. Analyzing the reaction using second order kinetics finds $k_f\ (a/V_b)$ falling between 2.03 - 4.06 (l/min cm) where the overall reaction system is represented by:

$$3\ CaBr_2 + H_2O \rightarrow (CaBr_2)_2{\cdot}CaO + 2\ HBr \quad (1)$$

This reaction is highly endothermic. At a design temperature of 1050 K (777°C), the heat of reaction is 181.5 kJ/mol (43.38 kcal/mol), and the Gibbs free energy change is positive at 99.6 kJ/mol (23.81 kcal/mol). Earlier work had envisioned a simpler hydrolysis system:

$$CaBr_2 + H_2O \rightarrow CaO + 2HBr \quad (2)$$

Here, hydrolysis is followed by Calcium bromide regeneration (≈850 K)

$$CaO + Br_2 \rightarrow CaBr_2 + \tfrac{1}{2}O_2 \quad (3)$$

The final hybrid stage is Hydrogen production and Bromine regeneration (≈493 K)

$$2HBr \rightarrow Br_2 + H_2 \quad (4)$$

The hydrolysis stage and is endothermic with an equilibrium constant of $5.39\times10^{-6}$ [3]. Therefore, means of supplying sufficient heat in the reaction process and shifting the reaction to the right to obtain adequate conversion must be identified. Two options are considered for bringing the reactants together continuously. In the first option, molten $CaBr_2$ is sprayed with or into a high temperature steam environment. The $CaBr_2$ droplets act as heat carriers for the reaction as well as one of the reactants. In the second option, steam bubbles are sparged through a pool of molten $CaBr_2$. Given the much greater heat carrying capacity of a pool of molten $CaBr_2$, the second option for the $CaBr_2$ reactor appears to be the more promising of the two and is being pursued as the primary candidate in the system design. Previous work on calcium bromide hydrolysis supported $CaBr_2$ on calcium-titanate substrates and ran at temperatures which were kept below the melting point of the $CaBr_2$ [4]. However this led to low yields and other technical barriers associated with the steam-solid-phase reaction.

An engineering design study at Argonne National Laboratory indicated that a sparging reactor concept can be readily scaled to commercial size and that the endothermic heat of reaction can be effectively supplied by heat transfer coils embedded in the melt. The experimental study focused on demonstrating that a sparging contactor can be used for $CaBr_2$ hydrolysis to develop an economical continuous process for hydrogen production and to obtain kinetic data. As indicated in Eqn. 1, this analysis shows that the hydrolysis reaction is complex and, therefore, a careful analysis of the kinetic data obtained with prototype reactor configuration is required. A systematic approach is used to determine the order of reaction in $CaBr_2$ and the reaction constant, which can be scaled-up to design multi-port steam sparger for a prototype hydrolysis reactor.

**Modeling results and discussion**

Argonne used the COMSOL™ [5] multi-physics partial differential equation (PDE) solver to investigate droplet spraying and bubble sparging molten calcium bromide reactor concepts, both operating above the melting point of $CaBr_2 \approx 1033$ K (760 °C) [6]. Examples of simulation results for the cases of laminar flow around a spherical drop of molten $CaBr_2$ and for a turbulent flow around a spherical-cap bubble of steam rising through a $CaBr_2$ melt are shown at the left and right of Fig. 1, respectively. The velocity in the axial direction is used to color the plots. The velocity scales just to the right of each plot are for the interior of the bubbles and to their right, for the external liquid flow fields. The plots also show streamlines and velocity vectors. The spherical bubble is near its maximum size as shown (Fig. 2). No recirculation zone appears in the wake, and therefore, the model indicates that reaction products would not be retained near the bottom surface of the bubble in its wake at significant concentrations. The flow pattern in the interior of the bubble shows that steam from the bubble interior near the centerline circulates to its surface where the reaction front would be for steam reacting with liquid $CaBr_2$.

The equivalent sphere diameter of the spherical-cap bubble is 4 mm with a Reynolds number of about 2500. The $k$-$\varepsilon$ turbulence model used [7] disguises the true nature of the wake, which in reality will consist of a chaotic mixture of vortices shedding from the bottom of the bubble. However, the model can be expected to capture the effect of entrainment of reaction products in the wake at the bottom of the bubble, which may affect reaction rates there. The interior of the cap also shows circulation that would bring steam from the interior to the reaction front at the bubble surface. Brennen [8], Clift, Grace, and Weber [9], and Wallis [10] have noted that contaminants in the liquid and surface tension and thermal gradients along the surface can act to slow this internal circulation.

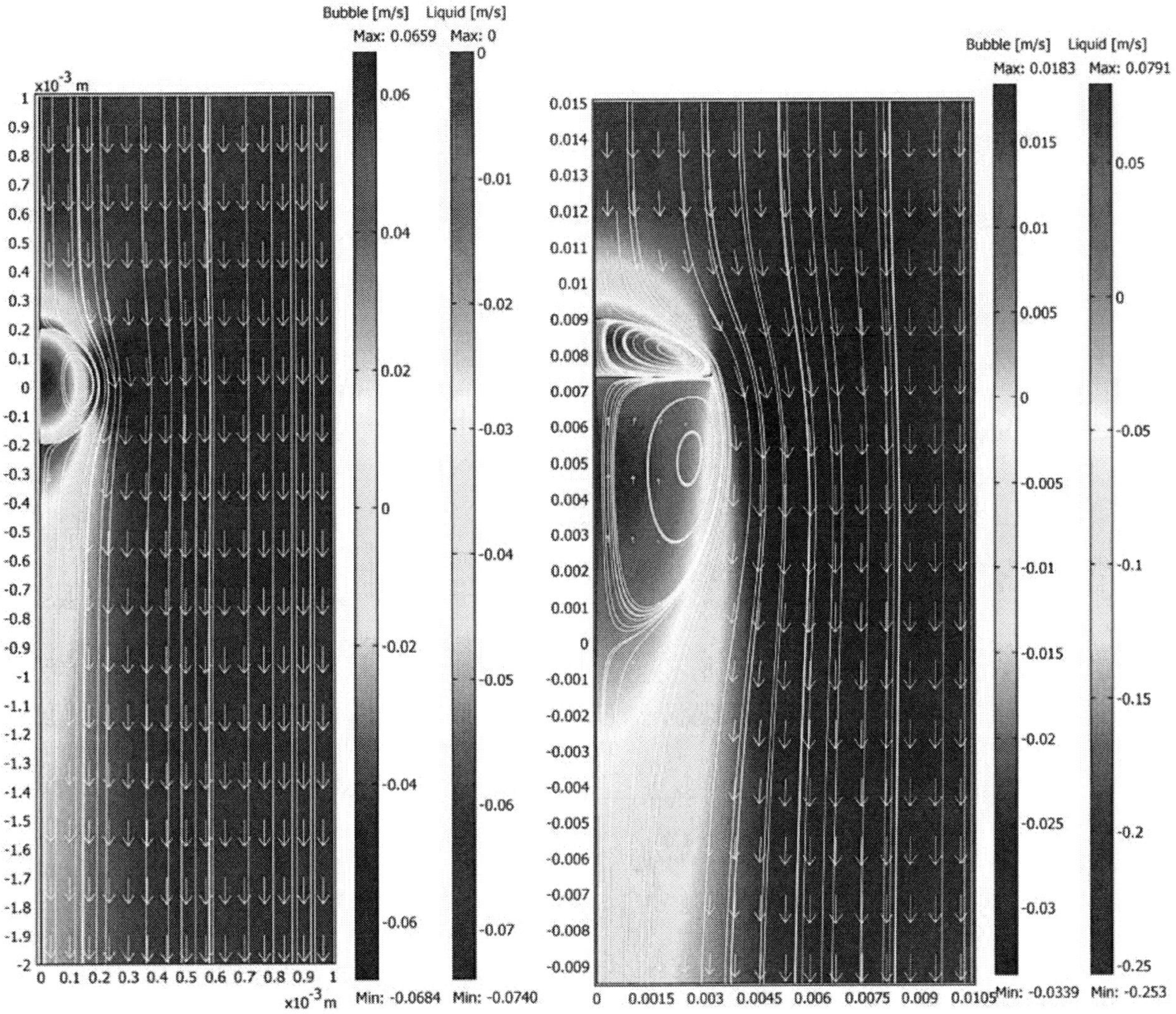


Figure 1. Left: Spray-dryer simulation of a laminar flow field around a 400 μm spherical $CaBr_2$ droplet moving at $u_\infty$= 6 cm/s. Right: Sparger simulation of the mean flow around a 4000 μm equivalent sphere diameter spherical-cap steam bubble rising at a velocity $u_\infty$= 20 cm/s in molten calcium bromide at 1013 K (740 °C).

A transient solution for a reacting spherical-cap bubble was performed using a reaction rate constant, $K = 10^{-8}$, producing a 0.47 HBr mole fraction in 1.0 s, the assumed residence time of a bubble in a typical bubble column. The initial temperature is 1023 K (750 °C). Plots of the HBr and CaO conversions and temperature field are shown in Figs. 3 and 4, respectively, at 1.0 s. The temperature scales to the right of the temperature plot are inside the bubble and in the external flow field, respectively. The temperature decrease due to the endothermic surface reaction was found to be negligible, producing an essentially isothermal temperature field within and in the vicinity of the bubble. This was anticipated because a 4 mm (4000 μm) equivalent sphere diameter bubble of steam can only extract heat from a very small volume of liquid calcium bromide. The concentration of CaO in the bubble wake is also found to be negligible. These results justify the steady flow field assumption.

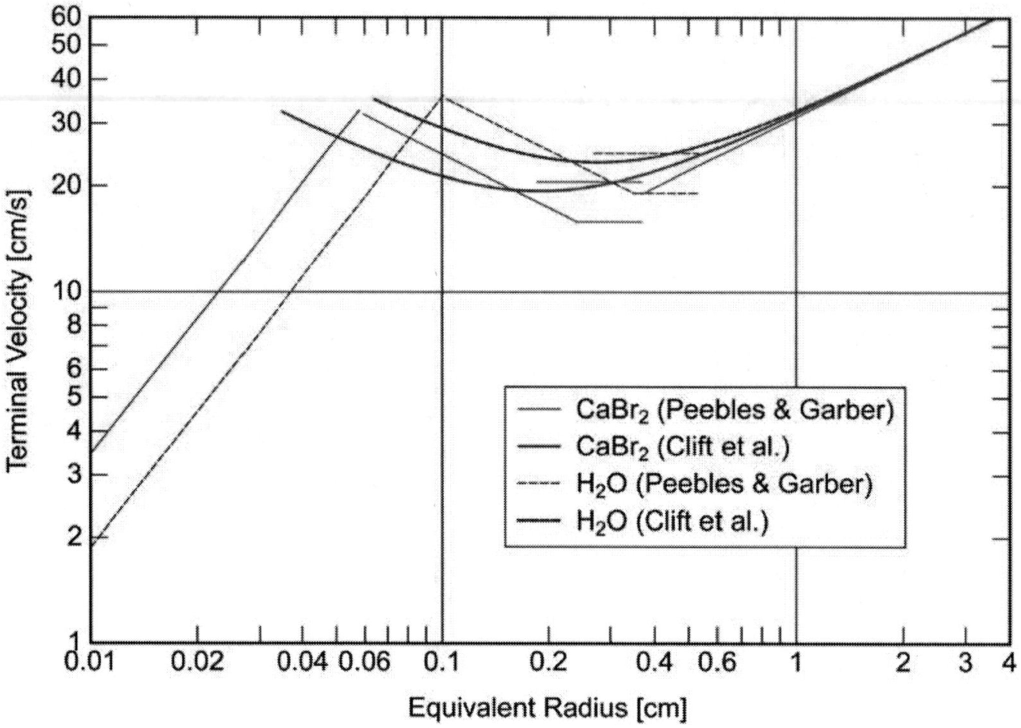


Figure 2. Terminal velocity of bubbles: water at 298 K (25 °C); calcium bromide at 1013 K (740 °C).

The relationship of this value of reaction rate constant to that analyzed from the Robinson et al. [11] data was determined. Robinson et al. conducted hydrolysis experiments at four temperatures by flowing steam over a porcelain boat placed in a silica tube heated electrically containing 0.5 g samples of solid calcium bromide crystals for 30 minutes. The weights of calcium bromide decomposed were reported but no analysis was performed. We analyzed these data as follows. Assuming an excess steam was used in these experiments, a first order reaction with calcium bromide was assumed to obtain reaction rate constants, $k^*$ for each temperature. The slope of the curve plotting the log of the resulting reaction rate constants vs. reciprocal of absolute temperature, T, were used to determine the energy of activation, E, in the Arrhenius equation given by;

$$k^* = A\,exp(-E/RT) \tag{4}$$

Then the pre-exponential, A, was evaluated. Using these values a reaction rate constant of 0.05 $min^{-1}$ at 1023 K was determined. In order to convert the reaction rate constant $k^*$ to $K$, the following expression was used:

$$K = k^*(V/S)/(C_{H_2O,i} \times 60) \tag{5}$$

where $C_{H_2O,i}$ is the initial molar concentration of steam (11.9 mol/m3) and the volume to surface area ratio of the equivalent spherical bubble V/S = r/3. Using an equivalent sphere radius, r = 2 mm, yields $K = 4.7\times10^{-8}$, which is on the same order of magnitude as the value of $10^{-8}$ used above. Using this value, the mole fraction of HBr at 1.0 s = 0.9. Analysis of the initial reaction rate data of Avogadro et al. [2] kinetics experiments for the HBr production and $CaBr_2$ consumption rates at 1023 K (750 °C) results in a reaction rate, $K$, in the range of approximately $0.5\text{-}1\times10^{-8}$. Note that these hydrolysis experiments used solid powdered calcium bromide. The HBr yields with values of k ranging from $10^{-7}$ to $4.7\times10^{-9}$ were computed for hydrolysis in molten calcium bromide. Table 1 summarizes the results of this parametric study. As before, the temperature fields were essentially isothermal.

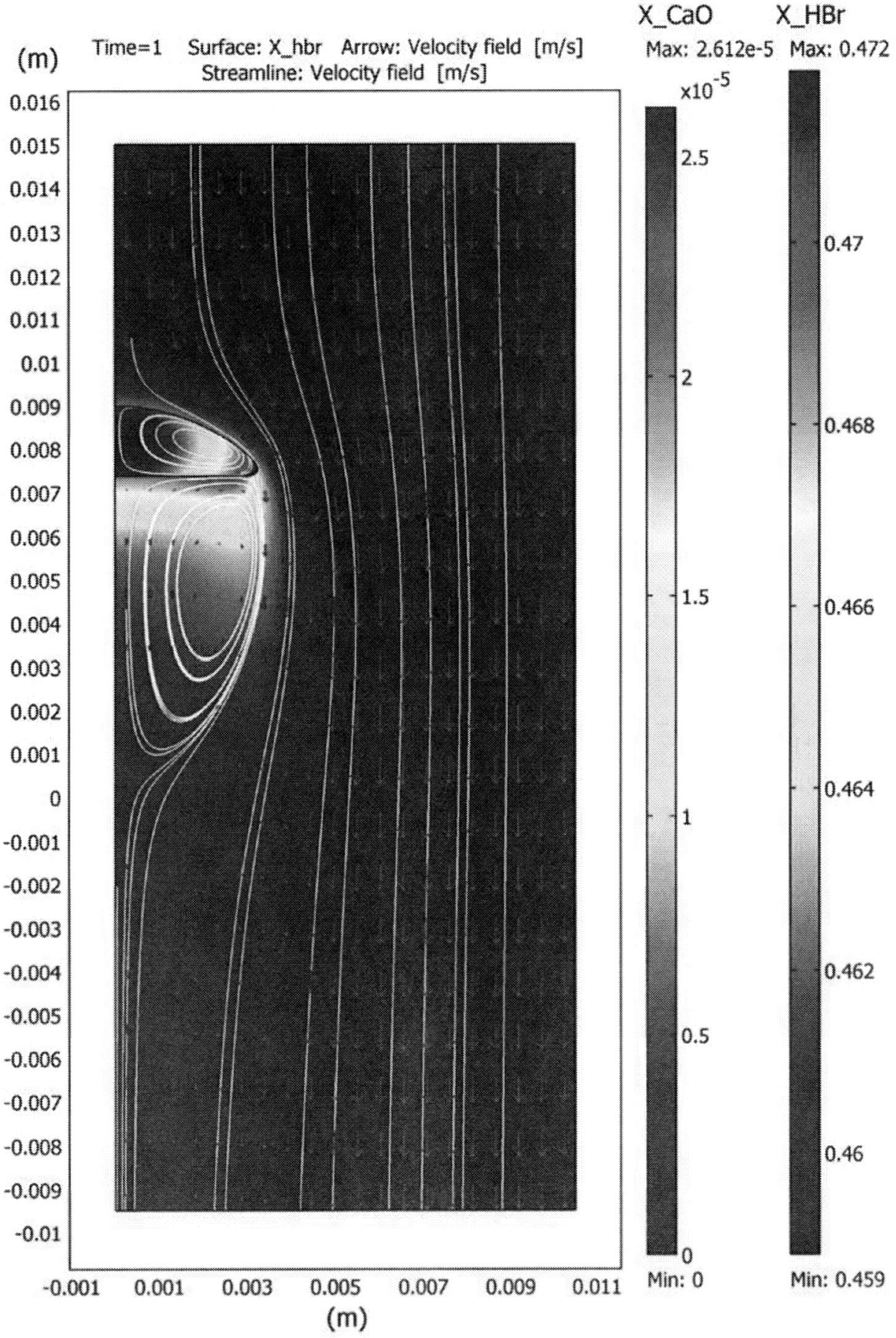


Figure 3. CaO and HBr mass fractions for a reacting 4 mm equivalent sphere diameter spherical-cap steam bubble rising at $u_\infty$= 20 cm/s in molten $CaBr_2$; Reaction rate constant, $K = 10^{-8}$, time = 1.0 s.

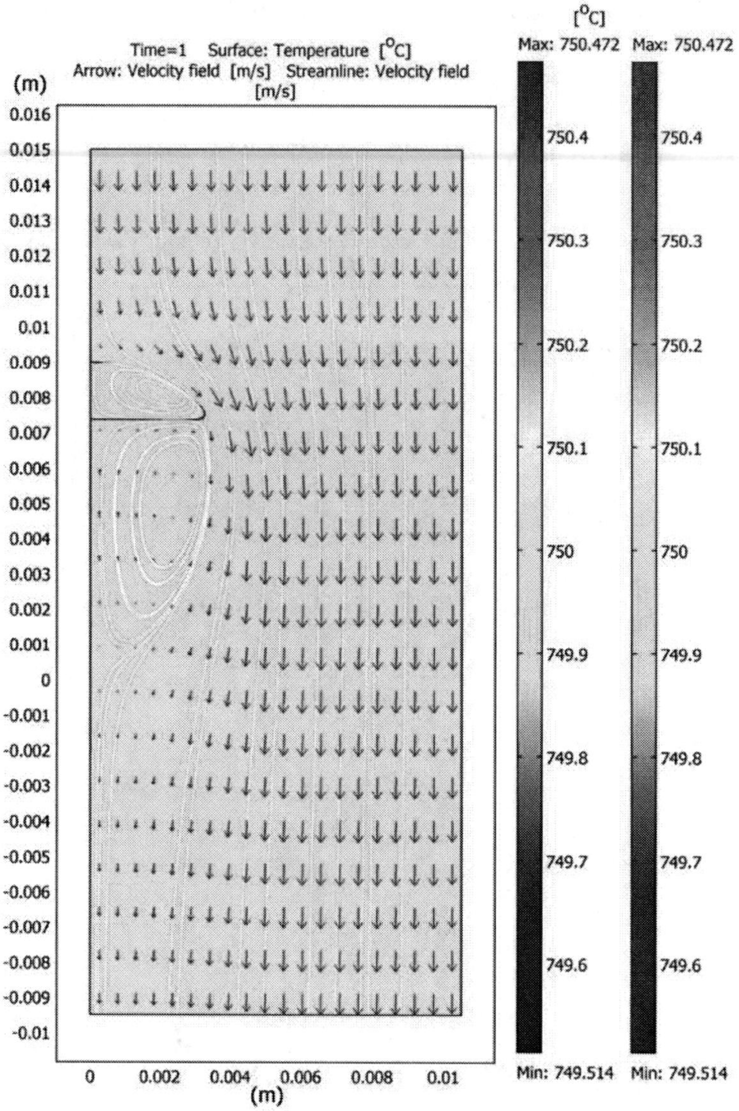


Figure 4. Temperature field for a reacting 4 mm equivalent sphere diameter spherical-cap steam bubble rising at 20 cm/s in molten calcium bromide. Reaction rate constant, $K = 10^{-8}$, time = 1.0 s.

Table 1. HBr production and steam conversion at 1.0 s as a function of reaction rate constant, k, at 1023 K (750 °C ).

| **Reaction Rate constant, *K*** | **Mole fraction HBr** | **Mole fraction steam** |
|---|---|---|
| $10^{-7}$ | 0.99+ | 0.01- |
| $4.7\times10^{-8}$ | 0.90 | 0.10 |
| $10^{-8}$ | 0.47 | 0.53 |
| $4.7\times10^{-9}$ | 0.28 | 0.72 |

Based on these modeling results it appeared that laboratory experiments involving the reaction of steam with molten calcium bromide in a sparging reactor would be successful, producing appreciable conversion. In fact, preliminary experiments at Argonne have produced reactions rates up to an order of magnitude higher than previous experiments [2, 11] using solid calcium bromide either as a matrix or as crystals.

The models developed in this paper, combined with further experiments to determine the reaction rate will help to size the molten $CaBr_2$ hydrolysis reactor and to identify the optimum operating range for bubble size to maximize conversion of steam to HBr in the reactor. The reaction might produce a eutectic and/or solid calcium oxide product which would be carried away by convection from the bubble surface reaction front. Some of these reaction products may end up in the wakes of bubbles, and the effects on the reaction rate at the trailing edge of the bubble surface need investigation. The bubble models will couple internal circulation, mixing, and diffusion of steam and HBr to and from the reaction front at the surface. Results of this modeling will aid in reactor design. Additional modeling may be required aid in sparger tube or bubble distributor design. The modeling should also be extended to include the effects of bubbles riding in the wakes of other bubbles. Based on results of bubble scale modeling, sub-models could be developed to characterize reaction progress and interfacial heat, mass, momentum transfer between bubbles and the surrounding molten $CaBr_2$ pool. These sub-models could then be integrated.

## Description of Argonne's Hydrolysis Experiments

The proof-of-concept $CaBr_2$ hydrolysis test reactor experiment apparatus briefly described in [3] operates in a batch mode. A schematic of the hydrolysis test reactor is shown in Fig. 5. The bench-scale reactor consists of a one-end-closed mullite tube (OD 76 mm, ID 70 mm, and 457 mm long) the other end of which is sealed with a stainless steel flange assembly. $CaBr_2$ melt is contained in a replaceable quartz or sintered alumina crucible (70 mm diameter by 200 mm long), which is placed inside the mullite reactor tube. The reactor tube is placed in a Lindberg high-temperature furnace with electric resistance heat elements. Steam is continuously sparged into the $CaBr_2$ melt through a 6.3 mm OD diameter tube (lance) made of SiC.

The product gases containing unreacted steam and HBr leave the reactor through a port in the stainless steel flange (or Pyrex glass flange) assembly. Then the product gases pass through a heat-traced line, a glass trap, a glass condenser chilled to ambient with cooling water and then be collected with (10 N) sodium hydroxide solution for HBr capture. The system is designed so that all gases introduced to it must pass through the condensate/neutralization tank where liquid or gaseous hydrogen bromide is neutralized.

The use of a concentrated sodium hydroxide solution (10 N) in sufficient volume (1.25 l) to convert all hydrogen bromide product formed into sodium bromide during the reaction time assures that even trace levels will not be released from the receiver to the vent.

As steam bubbles rise into the $CaBr_2$ melt, reaction occurs at the interface. The product gas HBr flows with steam and CaO goes into the melt solution. Samples of molten salts (calcium bromide-calcium oxide melts; ~2 mL) were collected on the exterior of an alumina probe thrust to the bottom of the molten salt bath. Melt samples are taken at the beginning of the test run before steam is turned on and at the end of test run after steam is shut off, and after purging with argon for a 5 minutes. The melt samples were also collected for analysis.

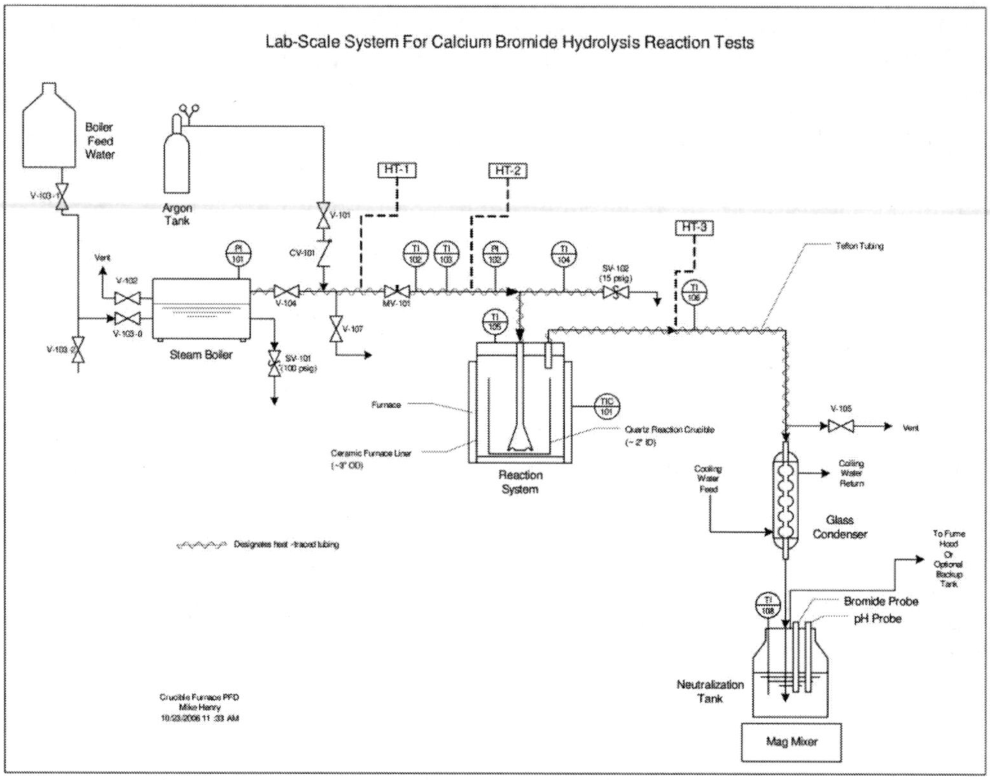


Figure 5. Schematic of $CaBr_2$ hydrolysis test apparatus
(Note: a straight sparger tube was used rather than funnel-type sparger shown).

During the contact time, $t_c$, of steam bubble with the $CaBr_2$ melt, the steam is consumed forming HBr. The contact time, $t_c$, is relatively small (~ 1 sec) as compared to the reaction time length (> 60 minutes) sampled in 10 to 20 minute intervals. Therefore, a pseudo-steady-state can be assumed for each bubble transit and the conversion of steam can be calculated assuming constant melt composition. Subsequently, the rate of $CaBr_2$ conversion can be determined by integrating the rate of steam conversion or HBr formation over the reaction time

## Reaction Model

Figure 6 shows the HBr formation rate as a function of reaction time. The HBr formation rate plot implies that the melt composition is changing as the reaction progresses. This could be explained on the basis of the reverse reaction occurring for the forward reaction given by Eqn. 2.

However, experiments were conducted with pure $CaBr_2$ and steam conversion (HBr formation) is low (<5%), the reverse reaction cannot explain the observed behavior. Furthermore, the phase diagram, shown in Figure 7, indicates that a $(CaBr_2)_2{\cdot}CaO$ complex is likely formed according to the reaction below:

$$2\,CaBr_2 + CaO \rightarrow (CaBr_2)_2{\cdot}CaO \qquad (6)$$

Note that there is an error in the original publication source [2] in Figure 7. The second entry in the abscissa should read $CaBr_2{\bullet}2CaO$.

This complex $(CaBr_2)_2{\cdot}CaO$ may not be reactive, effectively tying up a portion of the $CaBr_2$, making it unavailable for reaction with steam and may be even in the reverse reaction. The effective change of reactive $CaBr_2$ as the reaction proceeds would be more significant than $CaBr_2$ consumed by reaction alone. This implies that for each mole of CaO produced three moles of $CaBr_2$ are consumed; therefore, effective conversion of $CaBr_2$ could be limited to < 20 to 25%.

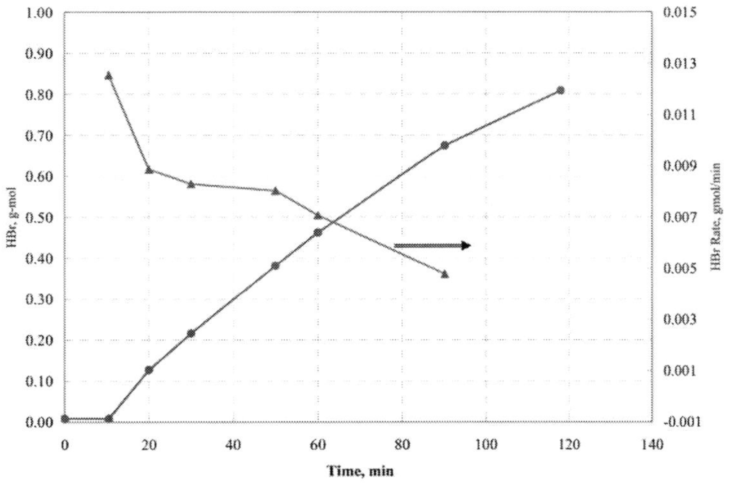


Figure 6. HBr formation rate as a function of reaction time (Test CBH-4).

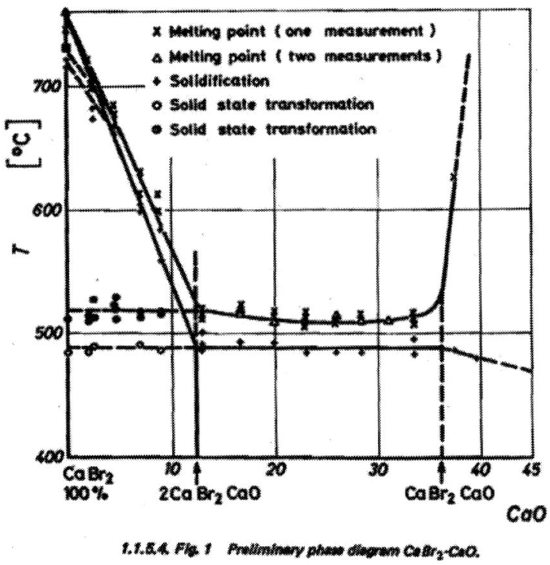


Figure 7. $CaBr_2$-CaO Eutectic phase diagram [source: Euratom, reference 2].

Combining the two reactions results in a net reaction as follows:

$$3\ CaBr_2 + H_2O \rightarrow (CaBr_2)_2{\cdot}CaO + 2\ HBr \quad (7)$$

Applying the reaction given by Equation (7) to a single bubble yields the following rate expression for HBr formation:

$$V_b \frac{d[HBr]}{dt_c} = k_f \bar{a}[P_{H_2O}][CaBr\ ]^n \quad (8)$$

The reaction rate of HBr is assumed to be first order in steam partial pressure and n[th] order (to be determined from the kinetic data) in $CaBr_2$ concentration. Similarly, the rate equation for consumption of $H_2O$ is given by:

$$V_b \frac{d[H_2O]}{dt_c} = -\frac{k_f}{2}\bar{a}[P_{H_2O}][CaBr\ ]^n \tag{9}$$

which can be simplified to the following form for integration over the contact time, $t_c$.

$$\frac{dP_{H_2O}}{dt_c} = -k'P_{H_2O} \tag{10}$$

Where:

$$k' = \frac{k_f(\bar{a}/V_b)}{2\rho_m}[CaBr_2]^n \tag{11}$$

Integration of Equation (2) yields the partial pressure of steam at a given contact time, $t_c$.

$$P_{H_2O} = P_{H_2O}(t_c = 0)e^{-k't_c} \tag{12}$$

The steam conversion $Y_{H_2O}$ and the rate of steam consumption $\dot{R}_{H_2O}$ are given by:

$$Y_{H_2O} = 1 - e^{-k't_c} \tag{13}$$

And:

$$\dot{R}_{H_2O} = \dot{m}_{H_2O}Y_{H_2O} \tag{14}$$

The rate of HBr formation (g-mole/min) is two times the rate of steam consumption. The rate of HBr formation varies with the reaction time $t_R$ as the melt composition changes.

In this analysis we applied 1st and 2nd order reaction in $CaBr_2$ to the baseline test run CBH-4 assuming its consumption is given by reactions Equation (2) (no complex) and Equation (7) (soluble complex formation), respectively. The root mean square (*RMS*) error, defined as,

$$RMS = 1/N\sqrt{\left[\sum\left(\dot{R}_e - \dot{R}_c\right)\Big/ R_e\right]^2} \tag{16}$$

where $\dot{R}_e$ and $\dot{R}_c$ are the experimental and calculated rates of HBr formation and *N* is the number of data points.

The *RMS* error values for 1st and 2nd order reactions in $CaBr_2$ consumption by reaction Equation (2) (no complex) were 0.0727 and 0.0486, respectively. Similarly, the *RMS* error values for 1st and 2nd order reactions based on $CaBr_2$ consumption by reaction Equation (7) (soluble complex formation) were 0.0193 and 0.0143, respectively. It was concluded that a 2nd order reaction in $CaBr_2$ represented by the reaction Equation (7) was found to match the experimentally far better than the other three cases. Figure 8 shows the comparison of calculated and experimental rates of reaction as a function of time. The derived reaction constants for test run CBH-4 and three other tests are shown in Table 2. It should be noted that the grouping $k_f(\bar{a}/V_b)$ is an effective reaction rate as commonly used in gas-liquid reactions [12], where $(\bar{a}/V_b)$ represents the interfacial area per unit reactor volume.

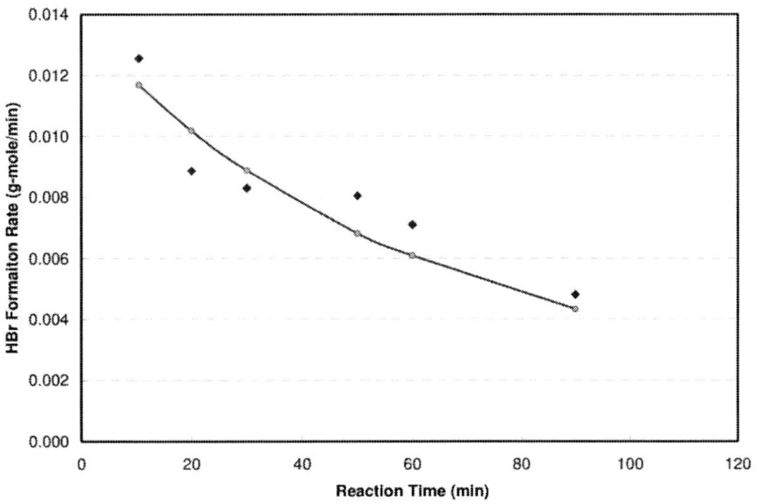


Figure 8. Comparison of predicted and experimental rates of HBr formation; 2nd order and $CaBr_2$ consumed by reaction Equation (8) for the soluble complex formation.

Table 2. Reaction constants of hydrolysis tests

| Test runs | Reaction Temperature, K (°C) | $CaBr_2$ melt height cm | Reaction constant, 1/(min cm) $k_f.\ (a/V_b)$ |
|---|---|---|---|
| CBH-2 | 1051.8 (778.6) | 4.5 | 2.03 |
| CBH-4 | 1046.1 (772.9) | 8.0 | 3.43 |
| CBH-5 | 1050.8 (777.6) | 4.5 (alumina beads) | 4.06 |
| CBH-10 | 1020.5 (747.3) | 8.0 | 1.28 |

A small difference in the combined reaction rate constant $k_f(\bar{a}/V_b)$ in tests CBH-4 and CBH-5 can be explained on the basis of temperature differences. However, the lower value in CBH-2 is mostly due to lower contact area per unit volume. Test run CBH-4 had an approximately 8 cm melt height, while test run CBH-2 had an approximately 4.5 cm melt height. By adding alumina beads to a shallow melt height, the contact area significantly improved as demonstrated by the higher reaction constant in test run CBH-5. The results show that by increasing the melt height, the distribution of steam improved. The steam distribution further improved with the sparger tube inserted into alumina beads at the bottom of the reactor crucible. The steam distributions for tests CBH-4 and CBH-10 were comparable; therefore, the reaction constant represents the Arrhenius rate constant. The activation energy derived from these two test runs is 341 kJ/mol, which is greater than the heat of reaction, 181 kJ/mol.

## Conclusions and Recommendations

- The modeling analysis indicates that good conversion of steam to HBr can be obtained over a broad range of kinetic rate constant, *K*, in a range determined from published experimental data [2,11]. In a pilot-scale unit, many more bubbles are required to achieve the desired production rate of HBr.
- The analysis of the Argonne experimental hydrolysis reaction data indicates that a second order hydrolysis reaction in $CaBr_2$ may be possible because of the formation of a soluble complex, hypothesized to be $nCaBr_2 \bullet CaO$.
- Formation of a complex, $nCaBr_2 \bullet CaO$, which may not be fully participating in the hydrolysis reaction, appears to explain high dependence of the rate of HBr formation on melt composition. The phase diagram given by Figure 3 indicates the value of $n = 2$.
- The activation energy, 341 kJ/mol, based on the analysis of the two test runs CBH-4 and CBH-10 is high compared to the heat of reaction, 181 kJ/mol. An additional test run at an intermediate temperature is required to validate this value of activation energy.
- If the basic conclusion of the existence of the soluble complex $nCaBr_2 \bullet CaO$ is correct, then it is essential to determine the reactivity of this complex in the regeneration reaction. A hydrolysis test followed by regeneration is strongly recommended. Premixed solution of $CaBr_2$ and CaO should not be used for the regeneration tests.

## Acknowledgements

This effort was sponsored by the U.S. Department of Energy's (DOE's) Nuclear Hydrogen Initiative: N-AN07TC0401. The investigations described herein were conducted by Argonne National Laboratory under DOE Contract No. DE-AC02-06CH11357.

## REFERENCES

[1] DOE Office of Nuclear Energy, *Nuclear hydrogen initiative: ten year program plan, Office of Advanced Nuclear Research*, Science and Technology, March 2005.

[2] Avogadro, A. et al., *Hydrogen production from water using nuclear heat*, Progress Report No. 3, Joint Nuclear Research Centre, Ispra Establishment, Italy, EUR 5059e ,1974.

[3] Doctor, R.D., Yang, J., Panchal, C.B., Lottes, S.A., and Lyczkowski, R.W., *Nuclear hydrogen initiative: proof of concept of the $CaBr_2$/CaO hydrolysis sparging contactor for HBr production from water-splitting*, Argonne National Laboratory Level-2 Milestone Report N-AN07TC0401, November 20, 2006.

[4] Nakayama T., "MASCOT-A bench-scale plant for producing hydrogen." International Journal of Hydrogen Energy, 1984, 9: 187-190.

[5] COMSOL, Inc., Burlington, MA, *COMSOL Version 3.2a User's Guide* (2004).

[6] Lottes S.A., Lyczkowski R.W., Panchal C.B., Doctor R.D., "Modeling and analysis of calcium bromide hydrolysis," Submitted to Int. J. of Hydrogen Energy 2007.

[7] Launder, B.E., Spalding, D.B., "The numerical computation of turbulent flows," Comp. Methods Appl. Mech. Eng. 1974; 3: 269–289.

[8] Brennen, C.E., Fundamentals of multiphase flows, Cambridge University Press (2005).

[9] Clift, R., Grace, J., Weber, M.E., Bubbles, drops, and particles, Academic Press, Inc. (1978).

[10] Wallis, G.B., One-dimensional two-phase flow, McGraw-Hill, Inc., New York (1969).

[11] Robinson, P.L., Smith, H.C., Briscoe, H.V.A., "The hydrolytic action of low-pressure superheated steam on salts of the alkaline earth metals," J. Chem. Soc., 1926; 129: 836-839.

[12] Danckwerts, P.V., Gas-Liquid Reactions, McGraw-Hill, Inc., New York (1970).

# AUTHOR INDEX